Jane's

DEFENCE GLOSSARY

First Edition

Compiled by
Ian Kay
Mary Walker

JANE'S INFORMATION GROUP

ISBN 0–7106–1118–8

Jane's Data Division
"Jane's" is a registered trade mark

163 Brighton Road, Coulsdon, Surrey CR5 2NH, United Kingdom

In the USA and its dependencies
Jane's Information Group Inc, 1340 Braddock Place
Suite 300, Alexandria, VA 22314–1651, USA

Printed in the United States of America

CONTENTS

Page

Introduction v

Main Glossary Listing 1

Appendices:

Officer ranks for NATO countries 251

JETDS coding system 256

US military aircraft designations 258

US missile & RPV designations 260

NATO reporting names for former Soviet aircraft 262

NATO reporting names for former Soviet missiles 264

Frequency bands & designations 267

Body armour threat levels 268

Units of measurement 269

Conversion factors 272

Periodic table 275

International phonetic alphabet 278

Company types & abbreviations 279

US states & abbreviations 281

Membership of international organizations 282

Country information 287

INTRODUCTION

Jane's Defence Glossary has been compiled as a companion to the wide range of other Jane's defence and aerospace publications. Available in either hard copy or on Jane's CD–ROM, it enables users to quickly identify over 11,000 defence–related acronyms and abbreviations along with their expanded descriptions.

Also included in the Glossary are a number of appendices providing a wealth of useful information on subjects ranging from NATO Officer Ranks to units of measurement and conversion factors.

Any comments on how to improve the usefulness of this Glossary or suggestions for additional information to be included in future editions would be welcomed and should be addressed to:

Ian Kay, Manager New Product Development
Tel: (+44 81) 763 1030
Fax: (+44 81) 763 1006

For details of how to obtain further copies of this Glossary or for information on other Jane's publications, please contact:

Jane's Information Group
Department DSM

Sentinel House
163 Brighton Road
Coulsdon, Surrey CR5 2NH
United Kingdom

Tel: (+44 81) 763 1030
Fax: (+44 81) 763 1006

1340 Braddock Place
Suite 300
Alexandria, VA 22314–1651
United States

Tel: (+1 703) 683 3700
Fax: (+1 703) 836 0029

A

A3F	Anglo-French Future Frigate
A3P	Advanced Avionic Architecture & Policy
AA	Active Adjunct (sonar)
AA	Air-to-Air
AA	Aluminum Association (USA)
AA	Anti-Aircraft
AA	Australian Army
AAA	Anti-Aircraft Artillery
AAA	Automatic Anti-Aircraft
AAAA	Army Aviation Association of America Inc.
AAAD	Airborne Anti-Armour Defence
AAAD	All Arms Air Defence (UK)
AAAI	American Association for Artificial Intelligence
AAAI	Association of Australian Aerospace Industries
AAAM	Advanced Air-to-Air Missile
AAAV	Advanced Amphibious Assault Vehicle
AAAW	Advanced Airborne Anti-Armour Weapon
AABM	Air-to-Air Battle Management
AABNCP	Advanced Airborne National Command Post (USA)
AAC	Alaskan Air Command (USA)
AAC	All-Aspect Capability
AAC	American Aerospace Controls Inc.
AAC	Anti-Aircraft, Common
AAC	Army Air Corps (UK)
AAC	Automatic Amplitude Control
AACC	Aircraft Accessories & Components Co (Saudi Arabia)
AACI	Advanced Applications Consultants Inc. (USA)
AACOMS	Army Area Communications System
AACP	Advanced Airborne Command Post
AACS	Airborne Advanced Communications System
AACT	Air-to-Air Combat Test (USA)
AACT	All Arms Classroom Trainer
AADEOS	Advanced Air Defense Electro-Optical Sensor
AADGE	Allied Command Europe Air Defence Ground Environment (NATO)
AAE	Aircraft Appliances & Equipment Ltd (Canada)
AAED	Advanced Airborne Expendable Decoy
AAF	Auxiliary Air Force (UK)
AAFCE	Allied Air Forces Central Europe (NATO)
AAFSS	Advanced Aerial Fire-Support System
AAG	Air-to-Air Gunnery
AAG	Anti-Aircraft Gun
AAGW	Air-to-Air Guided Weapon
AAH	Advance Attack Helicopter
AAHQ	Assembly Anchorage Headquarters (NATO)
AAI	American Aerospace Industries Inc.
AAI	Angle of Approach Indicator
AAICP	Air-to-Air Interrogator Control Panel

AAIP	Analogue Autoland Improvement Programme
AAK	Applique Armor Kit (USA)
AAL	Above Aerodrome Level
AALAAW	Advanced Air-Launched Anti-Armour Weapon
AALC	Amphibious Assault Landing Craft (USA)
AAM	Air-to-Air Missile
AAMI	Association for the Advancement of Medical Instrumentation (USA)
AAMTU	Automatic Antenna Matching & Tuning Unit
AANCP	Advanced Airborne National Command Post (USA)
AAO	Airborne Area of Operation
AAO	Air-to-Air Operations
AAOS	Acoustic Artillery Observer Sub-system
AAP	Anti-Aircraft, Practice
AAPSO	Afro-Asian Peoples' Solidarity Organisation
AAQMG	Assistant Adjutant & Quartermaster General (UK)
AAR	Air-to-Air Refuelling
AARA	Air-to-Air Refuelling Area
AARADCOM	Army Armament Research & Development Command
AARB	Advanced Aerial Refuelling Boom
AAS	Advanced Automation System
AAS	Agusta Aerospace Services SA (Belgium)
AAT	All Arms Trencher
AATCP	Air-Air Tres Courte Portee. Air-to-air very short range missile system (France)
AATH	Automatic Approach To Hover
AATS	Anti-Aircraft Training System
AATU	Automatic Antenna Tuning Unit (USA)
AAUS	Active Adjunct Undersea Surveillance
AAV	Arab American Vehicles Co. (Egypt)
AAV	Assault Amphibian Vehicle
AAV	Autonomous Aerial Vehicle
AAVP	Assault Amphibian Vehicle, Personnel
AAVR	Assault Amphibian Vehicle, Recovery
AAW	Advanced Analysts' Workstation
AAW	Anti-Air Warfare
AAWS	Advanced Anti-tank Weapon System
AAWS-H	Anti-Armor Weapons System – Heavy (USA)
AAWS-M	Advanced Anti-tank Weapon System – Medium
AAWWS	Airborne Adverse Weather Weapon System
AB	Able Rate (Royal Navy)
AB	Admiralty Board
AB	Air Base (USA)
AB	Airborne (forces)
AB	Assault Breaker
ABA	Adaption Box Assembly
ABACUS	Artillery Battery Computer Support system
ABAP	Association Belge pour l'Avion Patrouilleur (Belgium)
ABB	Asea Brown Boveri AB (Sweden)
ABC	All-purpose Ballastable Crawler
ABC	All-purpose Battle Computer

ABC	America, Britain & Canada
ABC	Argentina, Brazil, Chile
ABC	Atomic, Biological / Bacteriological & Chemical
ABC	Automatic Boost Control
ABC	Automatic Brightness Control
ABCA	America, Britain, Canada & Australia
ABCCC	Airborne Battlefield Command & Control Center (USA)
ABCD	American, British, Chinese & Dutch Allied Forces
ABD	Arab British Dynamics Co. (Egypt)
ABEB	Advanced Bridge Erection Boat
ABECO	Arab British Engine Co. (Egypt)
ABES	Air Breathing Engine System
ABF	Advanced Bomb Family (US Navy)
ABF	Annular Blast Fragmentation warhead
ABFM	American Board of Foreign Missions
ABH	Arab British Helicopters Co. (Egypt)
ABICS	ADA Based Integrated Control System
ABIS	All-Bus Instrumentation System
ABIT	Advanced Built-In Test
ABL	Airborne Laser
ABL	Anti-Blinde Leger. Light anti-armour (France)
ABL	Armoured Box Launcher
ABLE	Automotive Bridge Launching Equipment
ABM	Anti-Ballistic Missile
ABMD	Advanced Ballistic Missile Defence (USA)
ABMT	Anti Ballistic Missile Treaty
ABNL	Admiral, Benelux & Netherlands (NATO)
ABRC	Advisory Board for the Research Councils
ABRES	Advanced Ballistic Re-entry System
ABRI	Indonesian Armed Forces
ABRV	Advanced Ballistic Re-entry Vehicle
ABS	Aerospace Bearing Support Inc. (USA)
ABS (W)	Amphibious Bridging System (Wheeled)
ABSA	Advanced Base Support Aircraft
ABSP	Arab Ba'th Socialist Party
ABS(T)	Amphibious Bridging System (Tracked)
ABU	Aviation Bird Unit
AC	Active Component
AC	Aircraft Commander
AC	Analyst Console
AC	Area Coverage antenna
AC	Hydrogen Cyanide. Chemical warfare blood agent (also known as HCN)
ACA	Advanced Cargo Aircraft (US Army)
ACA	Agile Combat Aircraft
ACA	Allied Control Authority
ACA	Ammunition Container Assembly
ACAB	Air Cavalry Attack Brigade (USA)
ACADA	Automatic Chemical Agent Detector Alarm (USA)
ACAP	Advanced Composite Aircraft (helicopter) Programme

ACARD	Advisory Council for Applied Research & Development
ACARS	Automatic Communications And Reporting System
ACAS	Advisory, Conciliation & Arbitration Service
ACAS	Airborne Collision-Avoidance System
ACAS	Airfield Chemical Alarm System
ACAS	Assistant Chief of the Air Staff (UK)
ACAST	Advisory Committee on the Application of Science & Technology for Development
ACC	Air Combat Command (USAF)
ACC	Alternative Command Centre (NATO)
ACC	Area Control Centre
ACC	Artillery Control Console
ACC	Ateliers de Constructions du Centre (France)
ACC	Avionics Computer Control
ACC	Axis Controlled Carrier
ACCE	AFCEA Computing Conference & Exhibition
ACCHAN	Allied Command, Channel (NATO)
ACCIS	Automated Command & Control Information System (NATO)
ACCS	Advanced Communications Control System (USA)
ACCS	Air Command & Control System (NATO)
ACCS	Airborne Computing & Communications System
ACCS	Army Command & Control System (USA)
ACCSA	Allied Communications & Computer Security Agency (NATO)
ACCSCO	ACCS Company (NATO)
ACCU	Alternate Current Charging Unit (UK)
ACCV	Armored Cavalry Cannon Vehicle (USA)
ACD	Automated Chart Display
ACD	Automatic Call Distribution
ACDA	Arms Control & Disarmament Agency (USA)
ACDP	Armament Control & Display Panel
ACDS	Advanced Combat Direction System (USA)
ACDS	Advanced Countermeasures Dispenser System (USA)
ACDS	Air Combat Direction System
ACDS	Assistant Chief of Defence Staff
ACDS	Automatic Computer-controlled Dispensing System
ACDS	Automatic Countermeasures Dispensing System
ACE	Actuator Control Electronics
ACE	Agile Control Experiment
ACE	Alan Cobham Engineering Ltd (UK)
ACE	Allied Command Europe (NATO)
ACE	Analogue Code Encryption unit
ACE	Armored Combat Earthmover (USA)
ACE COMSEC	ACE Communications Security (NATO)
ACE High	ACE troposcatter communications system (NATO)
ACEATM	Aimed Controlled-Effect Anti-Tank Mine
ACEC	Atelier de Construction Electriques de Charleroi (Belgium)
ACELIP	ACE Long-term Infrastructure Plan (NATO)
ACEP	Automatic Contact Evaluation Plotter
ACES	Advanced Carry-on ELINT / ESM Suite
ACES	Advanced-Concept Escape System

ACES	Arrow Continuation Experiments
ACET	Air-Cushion Equipment Transporter
ACETEF	Air Combat Environment Test & Evaluation Facility (USA)
ACF	Army Cadet Force (British Army)
ACG	Amphibious Combat Group (Netherlands)
ACGF	Aluminium-Coated Glass-Fibre (chaff)
ACGS	Assistant Chief of the General Staff (UK)
ACI	American Concrete Institute
ACI	Armament Control Indicator
ACINT	Acoustic Intelligence
ACIS	Armament Control Indicator Set
ACK	Acknowledgement
ACLANT	Allied Command, Atlantic (NATO)
ACLICS	Airborne Communications Location Identification & Collection System (USA)
ACLOS	Automatic-to-Command Line-Of-Sight
ACLS	Air Cushion Landing System
ACLS	Automatic Carrier Landing System
ACLT	Aircraft Carrier Landing Trainer
ACM	Advanced Cruise Missile
ACM	Air Chief Marshall
ACM	Air Combat Manoeuvre
ACM	Anti-Armour Cluster Munitions
ACM	Australian Chamber of Manufacturers
ACME	Advanced-Core Military Engine
ACMEST	Air Combat Manoeuvering Expert Systems Trainer
ACMI	Air Combat Manoeuvring Instrumentation
ACMR	Air Combat Manoeuvring Range
ACMS	Air Combat Manoeuvring Simulator
ACMS	Aircraft Condition Monitoring System
ACMS	Armament Control & Monitoring System
ACMS	Automatic Communication Management System
ACMT	Advanced Cruise Missile Technology
ACN	Aircraft Classification Number
ACNIP	Auxiliary Communications, Navigation & Identification Panel
ACNS	Assistant Chief of the Naval Staff (UK)
ACO	Administrative Contracting Officer
ACO	Airspace Control Order
ACOS	Assistant Chief Of Staff
ACOST	Advisory Committee on Science & Technology
ACOVS	Artillery Combat Observer Vehicle Simulator
ACO(W)	Atomic Co-ordinating Office (Washington)
ACP	African / Caribbean / Pacific (NATO)
ACP	Airborne Command Post
ACP	Allied Communications Procedures / Publication (NATO)
ACP	Altimeter Check Point
ACP	Armament Control Panel
ACP	Automatic Colt Pistol
ACPA	Adaptive-Controlled Phased Array
ACPM	Anti-Char a Pose Mecanique (France)

ACPR	Anti-Char a Pose Rapide. Anti-tank mine (France)
ACR	Advanced Combat Rifle
ACR	Aerial Combat Reconnaissance
ACR	Air Control Radar
ACRIM	Active Cavity Radiometer Irradiance Monitor
ACRV	Armoured Command & Reconnaissance Vehicle
ACs	Agreed Characteristics
ACS	Advanced Communications Systems Inc. (USA)
ACS	Air Combat Simulator
ACS	Air Commando Squadron (USAF)
ACS	Airborne Control System
ACS	Application Control Structure
ACS	Armament Control System
ACS	Attitude Control System
ACS	Automatic Channel Selection
ACSA	Allied Communications Security Agency
ACSE	Association Control Service Element (NATO)
ACSM	Advanced Conventional Stand-off Missile
ACSM	Assemblies, Components, Spare parts & Materials (NATO)
ACSSB	Amplitude Companded Single Sideband
ACT	Active Control Technology
ACT	Airborne Crew Trainer
ACT	Airportable Cargo Trailer
ACT	Anti-Communications Threat
ACTEW	Acoustic Charged Transport Electronic Warfare. Low-cost decoy system
ACTIS	Advanced Compact Thermal Imaging System
ACTS	Advanced Communications Technology Satellite (NASA)
ACU	Acceleration Control Unit
ACU	Adaptive Control Unit
ACU	Airborne Computer Unit
ACU	Analysis Control Unit
ACU	Artillery Computer Unit
ACU	Avionics Control Unit
ACV	Acquisition & Command Vehicle
ACV	Air Cushion Vehicle
ACV	Armored Cannon Vehicle (USA)
ACV	Armoured Combat Vehicle
ACV	Armoured Command Vehicle
ACVT	Armored Combat Vehicle Technology Program (USA)
ACVT	Armoured Combat Vehicle, Tracked
ACWAR	Agile Continuous Wave Acquisition Radar
ACWS	Advanced Color Work Station (USA)
AC/RC	Active Component / Reserve Component
AD	Accidental Damage
AD	Air Defence
AD	Air Division (USAAF, USAF)
AD	Analog / Digital
AD	Area-Denial munition
AD	Armament Division (AFSC) (USA)

AD	Assistant Director (UK)
AD	Destroyer Tender (USA)
ADA	Action Data Automation
ADA	Air Defence Artillery (USA)
ADA	Air Defended Area
ADAC	Appareil de Detection d'Alerte Chimique. Chemical warning device (France)
ADAC	Auto Directeur Anti-Clutter. Active radar seeker (France)
ADACS	Airborne Digital Automatic Collection System
ADACS	Alarm & Distributed Access Control System
ADAD	Air Defence Alerting Device
ADAIRS	Air Data And Inertial Reference System
ADAM	Area Denial Artillery Munition
ADAMS	Air Defence Anti-Missile System
ADAMS	Air Defense Advanced Mobile System
ADAPS	Automatic Data Acquisition & Processing
ADARIO	Analog / Digital Adaptable Recorder Input / Output
ADARS	Adaptive Antenna Receiver System
ADAS	Airborne Data Acquisition System
ADAS	Airfield Damage Assessment System (USAF)
ADAS	Auxiliary Data Annotation Set
ADatP	Allied Data Publications (NATO)
ADat-P3	Automatic Data Processing (Standard) 3 (NATO)
ADATS	Air Defence Anti-Tank System (Switzerland)
ADAU	Auxiliary Data Acquisition Unit
ADAWS	Action Data Automation Weapons System (UK)
ADAZ	Air Defence Zone
ADC	Acoustic Device Countermeasure
ADC	Air Data Computer
ADC	Air Defense Command (Japan)
ADC	Air Defense Command (former USAF Command)
ADC	Analogue to Digital Converter
ADC	Armament Directors Conference
ADC	Asynchronous Digital Combiner
ADCA	Advanced Design Composite Aircraft (USAF)
ADCAP	Advanced Capabilities
ADCAPARM	Advanced Capabilities Anti-Radar Missile
ADCC	Air Defence Control Centre
ADCC	Air Direction Control Centre
ADCIS	Air Defence Command Information System (UK)
ADCOM	Aerospace Defence Command
ADCOR	American Development Corp.
ADCS	Air Data Computer System
ADC(X)	Auxiliary dry cargo carrier (USA)
ADD	Agency for Defence Development (Korea, South)
ADDC	Air Defence Data Centre (UK)
ADDS	Advanced Decoy Dispenser System
ADDS	Advanced Digital Dispensing System
ADDS	Army Data Distribution System (USA)
ADE	Aeronautical Development Establishment (India)

ADE	Armoured Division Equivalent
ADEA	Army Development & Employment Agency (USA)
ADECS	Advanced Digital Engine Control System
ADEHML	Auto Diesels Edghill HML Ltd (UK)
ADELA	Atlantic Development Group for Latin America
ADELE	Alerte Detection Et Localisation des Emmetteurs. Airborne ESM system (France)
ADELT	Automatically Deployable Emergency Locator Transponder
ADEWS	Air Defense Electronic Warfare System (USA)
ADF	Automatic Direction Finder
ADFCC	Air Force Defence Control Centre (Japan)
ADG	Accessory-Drive Generator
ADG	Aircraft Delivery Group (USA)
ADGE	Air Defence Ground Environment
ADGS	Air Defence Gun Sight
ADHM	Aerospace Design Hugo Marom Ltd (Israel)
ADI	Air Decoy Missile (USA)
ADI	Air Defense Initiative (USA)
ADI	Attitude Director Indicator
ADI	Australian Defence Industries Ltd
ADIB	Air Deployable Ice Beacon
ADIMP	Ada Improvement Programme (UK)
ADIRS	Air Data & Inertial Reference System
ADIRS	Airfield Damage Information & Reporting System
ADIT	Automatic Detection & Tracking unit (USA)
ADITT	Aerially Deployed Ice Thickness Transponder
ADIVS	Air Defence Interoperability Validation System
ADIZ	Air Defence Identification Zone
Adjt	Adjutant
Adjt-Gen	Adjutant-General
ADKEM	Advanced Kinetic Energy Missile
ADL	Authorised Data List
ADLER	Artillerie-Daten-Lage und Einsatz-Rechnerverbund (Germany)
ADLIPS	Automatic Data Link Plotting System (USA)
Adm	Admiral
ADM	Advanced Development Model
ADM	African Democratic Movement (Ciskei)
ADM	Air Decoy Missile
ADM	Asynchronous Data Modem
ADM	Automatic Demolition Munition
Adm of Fleet	Admiral of the Fleet
Admin	Administration
ADMU	Air Distance Measuring Unit
ADNC	Air Defence Notification Centre
ADOC	Air Defence Operations Centre
ADOCS	Advanced Digital Optical Control System
ADOP	Advanced Distributed Onboard Processor
ADP	Arab Democratic Party
ADP	Automatic / Advanced Data Processing
ADPA	American Defense Preparedness Association

ADPCM	Adaptive Differential Pulse Code Modulation
ADPE	Automated Data Processing Equipment (USA)
ADPG	Air Defence Planning Group
ADPSO	Automatic Data Processing Selection Office (USA)
ADR	Accident Data Recorder
ADR	Air Data Relay
ADR	Air Defence Region (UK)
ADR	Airfield Damage Repair
ADRC	Advanced Defence Radar Controller
ADRG	ARC Digital Raster Graphics
ADROWPU	Advanced Double Pass Reverse Osmosis Water Purification Unit
ADRS	Airfield Damage Repair Squadron (Royal Air Force)
ADRTIES	Advanced Digital Radar Imagery Exploitation System
ADRU	Advanced Defence Radar Unit
ADS	Acoustic Detection System
ADS	Advanced Deployable Systems program (US Navy)
ADS	Air Data System
ADS	Ammunition Delivery System
ADS	Architectural Design Study (NATO)
ADS	Audio Distribution System
ADS	Automatic Defence System
AdSAAMS	Advanced Surface-to-Air Missile System
ADSAF	Automatic Data System for the Army in the Field (USA)
ADSCOM	Advanced Ship Communications (USA)
ADSI	Aerospace & Defense Sales Inc. (Philippines)
ADSIA	Allied Data Systems Interoperability Agency
ADSM	Air Defence Suppression Missile
ADSP	Advanced Digital Signal Processor
ADSU	Air Data Sensor Unit
ADT	Advanced Display Terminal
ADT	Air Data Terminal
ADT	Air Defence Technology programme
ADT	Airborne Data Terminal
ADT	Alphanumeric Display Terminal
ADT	Automatic Data Transmission
ADT	Automatic Detection & Tracking
ADT3	Air Defence Tactical Training Theatre
ADTC	Armament Development & Test Center (USAF)
ADTU	Auxiliary Data Transfer Unit
ADU	Air Data Unit
ADU	Annotation Display Unit
ADU	Auxiliary Display Unit
ADU	Avionics Display Unit
ADV	Air Defence Variant
ADVCAP	Advanced Capability
ADVON	Advanced Echelon
ADW	Area Denial Weapon
AD/XJAM	Artillery Delivered Expendable Jammer
AE	Ammunition ship (USA)

AE	Assault Echelon
AEA	Atomic Energy Authority
AEB	Active Electronic Buoy
AEC	Associated Equipment Company
AEC	Atomic Energy Commission (USA)
AECB	Arms Export Control Board
AECM	Active Electronic Countermeasures
AECMA	Association Europeenne des Constructeurs de Materiel Aerospatial (France)
AECU	Advanced Electronic Control Unit
AEDC	Arnold Engineering Development Center (USA)
AEDF	Asian Economic Development Fund
AEDS	Atomic Energy Detection System
AEEMA	Australian Electrical & Electronic Manufacturers Association Ltd
AEFS	Fleet replenishment ship
AEFSS	Automatic Explosion / Fire Sensing & Suppression system
AEGIS	Advanced Engine / Gearbox Integrated System
AEGIS	Airborne Early Warning Ground Integration Segment
AEHM	Airport Engineering Hugo Marom Ltd (Israel)
AEIA	Australian Electronics Industry Association
AEM	Applied Electro Mechanics Inc. (USA)
AEM	Aviation Engineering & Maintenance (UK)
AEMA	Artefatos Electricos e Mecanicos de Aeronautica Ltda (Brazil)
AEMC	Advanced Energetic Materials Corp. (USA)
AEO	Air Electronics Officer (Royal Air Force)
AEOS	Aircraft Equipment Overhauls & Sales (NSW) Pty Ltd (Australia)
AERCO	Aviation Electrical & Radio Co. Ltd (UK)
AERE	Atomic Energy Research Establishment
AERS	Access / Egress Roadway System
AES	Auger Electron Spectroscopy
AES	A. Eickhorn GmbH & Co. fur Schneidwaren & Waffen (Germany)
AESN	Alenia Elsag Sistemi Navali (Italy)
AETMS	Airborne Electronic Terrain Mapping System
AEU	Airborne Electronics Unit
AEV	Armoured Engineer Vehicle
AEW	Airborne Early Warning
AEWTF	Aircrew Electronic Warfare Tactics Facility (NATO)
AEW&C	Airborne Early Warning & Control
AF	Air Force
AF	Airfield (NATO)
AF	Amphibious Force (NATO)
AF	Audio Frequency
AF	Store Ship (USA)
AF RE	Chief of Air Force Reserve's Office (USAF)
AFA	Air Force Association (USA)
AFA	Audio Frequency Amplifier
AFAAC	Air Force & Anti-Aircraft Command (Switzerland)
AFAALC	Air Force & Anti-Aircraft Logistics Command (Switzerland)
AFAATC	Air Force & Anti-Aircraft Training Command (Switzerland)
AFAP	Artillery-Fired Atomic Projectile

AFARMADE Asociacion Espanola de Fabricantes de Armamento y Material de Defensa (Spain)
AFARV Armoured Forward Area Re-arm Vehicle
AFAS Advanced Field Artillery System
AFASD Air Force Aeronautics Systems Division (USA)
AFASPO Air Force Automated Systems Project Office (USAF)
AFATDS Advanced / Army Field Artillery Tactical Data System
AFATL Air Force Armaments Laboratory
AFB Air Force Base (USA)
AFBMA Anti-Friction Bearing Manufacturers Association (USA)
AFBMD Air Force Ballistic Missile Division (USA)
AFBMO Air Force Ballistic Missile Office (USA)
AFC Automatic Frequency Control
AFCAS Automatic Flight Control Augmentation System
AFCC Air Force Communications Command (USA)
AFCC Office of the Chief of Staff, USAF
AFCEA Armed Forces Communications & Electronics Association (USA)
AFCENT Allied Forces, Central Europe (NATO)
AFCMD Air Force Contract Management Division (USA)
AFCS Artillery Fire-Control System
AFCS Automatic Fire Control System
AFCS Automatic Flight Control System
AFCSA/CC Office of the Commander, Air Force Center for Studies & Analyses (USAF)
AFD Automatic Feeding Device (for ammunition)
AFDB Large auxiliary floating dry dock (USA)
AFDE Arctic Fuels Dispensing Equipment
AFDL Small auxiliary floating dry dock (USA)
AFDM Medium auxiliary floating dry dock (USA)
AFDS Advanced Flight-Deck Simulator
AFDS Air Fighting Development Squadron (UK)
AFDS Automatic Flight Director System
AFE Accessories for Electronics Inc. (USA)
AFEDDS Automatic Fire & Explosion Detection & Suppression System (Israel)
AFES Automatic Fire Extinguishing System
AFESD Air Force Electronic Systems Division (USA)
AFEWC Air Force Electronic Warfare Center (USA)
AFEWES Air Force Electronic Warfare Evaluation Simulator (USA)
AFEX Anglo-French Exchange meeting
AFF Ammunition Factory Footscray (Australia)
AFFF Aqueous Film-Forming Foam (fire fighting)
AFFTC Air Force Flight Test Center (USA)
AFG Anti-Frogman Grenade
AFGL Air Force Geophysics Laboratory (USA)
AFGS Automatic Flight Guidance System
AFIC Air Force Intelligence Command (USA)
AFIS Air Force Intelligence Service (USA)
AFLC Air Force Logistics Command (USA)
AFM Air Force Manual (USA)
AFMC Air Force Materiel Command (USA)
AFMED Allied Forces Mediterranean. Now NAVSOUTH (NATO)

AFMS	Automatic Flight Management System
AFNE	Astilleros y Fabricas Navales del Estado (Argentina)
AFNON	Allied Forces, North Norway (NATO)
AFNOR	Association Francaise de Normalisation
AFNORTH	Allied Forces, Northern Europe. Now replaced by AFNORTHWEST (NATO)
AFNORTHWEST	Allied Forces, Northwest Europe. The new major subordinate Command under ACE, replacing AFNORTH (NATO)
AFO	Area Flag Officer
AFOLTS	Automatic Fire Overheat Logic Test System
AFOSR	Air Force Office of Scientific Research (USA)
AFP	Arbeitsgemeinschaft Flughafen Planung (Germany)
AFP	Armed Forces of the Philippines
AFP	Assault Fire Platoon
AFPRO	Air Force Plant Representative Office
AFP+	Assault Fire Platoon Plus
AFR	Air Force Regulation
AFRES	Air Force Reserve (USA)
AFRP	Aramid Fibre Reinforced Plastic
AFS	Air Force Station (USA)
AFS	Automatic Flight System
AFS	Automatic Frequency Selection
AFS	Aviation Fluids Service Inc. (USA)
AFS	Combat Stores Ship (USA)
AFSARC	Air Force Systems Acquisition Review Council
AFSAT	Air Force Satellite
AFSATCOM	Air Force Satellite Communications System (USA)
AFSC	Air Force Systems Command (USA)
AFSCF	Air Force Satellite Control Facility (USA)
AFSD	Air Force Space Division (USA)
AFSED	Arab Fund for Economic & Social Development
AFSK	Audio Frequency Shift Keying
AFSONOR	Allied Forces, South Norway (NATO)
AFSOUTH	Allied Forces, Southern Europe (NATO)
AFSTC	Air Force Space Technology Center (USA)
AFSV	Armoured Fire Support Vehicle
AFT	Air Force Department (NATO)
AFTA	Advanced First Term Avionics
AFTI	Advanced Fighter Technology Integration
AFTS	Airtech Firearms Training System
AFU	Air Filtration Unit
AFU	Assault Fire Unit
AFV	Armored Family of Vehicles (USA)
AFV	Armoured Fighting Vehicle
AFWAL	Air Force Wright Aeronautical Laboratory (USA)
AFWL	Air Force Weapons Laboratory (USA)
AF/IG	Inspector General's Office (USAF)
AF/IN	Office of the Assistant Chief of Staff, Intelligence (USAF)
AF/SC	Office of the Deputy Chief of Staff, Command, Control, Communications & Computers (USA)

AG	Adjutant-General (UK)
AG	Air Gunner
AG	Antenna Group
AG	Army Garrison (US Army)
AG	Miscellaneous ship (USA)
AGARD	Advisory Group for Aerospace Research & Development
AGATHA	Air / Ground Anti-jam Transmission from Helicopter or Aircraft (France)
AGC	Adjutant-General's Corps (UK)
AGC	Automatic Gain Control
AGCS	Advanced Guidance & Control System
AGCW	Autonomously Guided Conventional Weapon
AGDS	Deep submergence support ship (USA)
AGE	Advanced Graphics Engine
AGE	Algram Engineering Co. Ltd (UK)
AGEH	Hydrofoil research ship (USA)
AGES/AD	Air-to-Ground Engagement / Air Defence
AGETS	Automated Ground Engine Test System
AGEX	Anglo-German Exchange meeting
AGF	Army Ground Forces
AGF	Miscellaneous command ship (USA)
AGFF	Frigate research ship (USA)
AGI	Aeronautical & General Instruments Ltd (UK)
AGI	Intelligence Gathering Auxiliary vessel
AGIS	Assieme per Gestione e Impiegodel Sistema. FSAF command system (Italy)
AGIS	Automatisiertes Gefechts – und Informationssystem fur Schnellboote (Germany)
AGL	Above Ground Level
AGL	Airborne Gun Laying (radar)
AGL	Automatic Grenade Launcher
AGLS	Automatic Gun Laying System
AGLT	Airborne Gun Laying Turret
AGM	Air-to-Ground Missile
AGM	Missile range instrumentation ship (USA)
AGMA	American Gear Manufacturers Association
AGOR	Oceanographic Research ship (USA)
AGOS	Ocean Surveillance ship (USA)
AGP	Patrol craft tender (USA)
AGPPE	Advanced General Purpose Processor Element
AGR	Advanced Gas-cooled Reactor
AGR	Radar picket ship
AGRA	Automatic Gain Ranging Amplifier
AGREE	Advisory Group on Reliability in Electronic Equipment
AGS	Aeronautical Ground Station
AGS	Armored Gun System (USA)
AGS	Assault Gun System vehicle (USA)
AGS	Surveying Ship (USA)
AGSE	Advanced Ground Systems Engineering Corp. (USA)
AGSS	Auxiliary research submarine (USA)

AGTS	Air Gunnery Target System
AGV	Assault Gun Vehicle
AGV	Automated Guided Vehicle
AGVT	Advanced Ground Vehicle Technology
AGZ	Actual Ground Zero
AG/FF	Frigate / FAC support ship
AH	Artificial Horizon
AH	Hospital ship (USA)
AHCP	Ad Hoc Collective Protection
AHE	Ammunition Handling Equipment
AHEAD	Advanced Hit Efficiency And Destruction
AHEM	Association of Hydraulic Equipment Manufacturers
AHG	Ateliers de la Haute Garonne Ets Auriol & Cie (France)
AHIP	Army Helicopter Improvement Program (USA)
AHIT	Advanced Hover Intercept Technology
AHM	Anti-Helicopter Mine
AHQ	Air Headquarters
AHQ	Allied Headquarters (NATO)
AHRS	Attitude & Heading Reference System
AHRU	Attitude & Heading Reference Unit
AHS	American Helicopter Society
AI	Air Interdiction
AI	Airborne Intercept
AI	Alpha Industries Inc. (USA)
AI	Artificial Intelligence
AI	Automatic Identification
AI2	Avionic Instruments Inc. (USA)
AI2S	Advanced Infra-red Imaging Seeker
AIA	Aerospace Industries Association of America
AIA	Associazione Industrie Aerospaziali (Italy)
AIAA	American Institute of Aeronautics & Astronautics
AIAAM	Advanced Intercept Air-to-Air Missile
AIAG	Automotive Industries Action Group
AIAIT	Army Intelligence & Threat Analysis Center (US Army)
AIBU	Advanced Interference Blanker Unit
AICORN	Adaptive Intelligent Control of Radio Networks
AICRO	Association of Independent Contract Research Organisations
AID	Agency for International Development (USA)
AIDB	Anti-Intrusion Defence Barriers
AIDC	Aero Industry Development Center (Taiwan)
AIDS	Acoustic Intelligence Data System
AIDS	Advanced Integrated Display System (USA)
AIDS	Airborne Integrated Data System
AIECO	Ahmed International Equipment Co. (Pvt) Ltd (Pakistan)
AIEWS	Advanced Integrated Electronic Warfare System
AIF	Air International Formation SA (France)
AIFF	Advanced Identification Friend or Foe
AIFS	Advanced Indirect Fire System
AIFTA	Anglo-Irish Free Trade Agreement
AIFV	Armored Infantry Fighting Vehicle (USA)

AIG	Address Indicator Group
AIG	Assistant Instructor of Gunnery (Royal Air Force)
AIIM	Association for Information & Image Management (USA)
AIL	Advanced Information Leaflets
AILA	Airborne Instrument Landing Approach system
AILC	AMRAAM International Licensing Company
AIM	Air Intercept Missile
AIM	Air traffic control beacon IFF Mk XII system
AIM	Aircraft Interphone & Mixer (Yugoslavia)
AIM	Anti-Invasion Mine
AIMES	Avionics Integrated Maintenance Expert System
AIMIS	Advanced Integrated Modular Instrumentation System (US Navy)
AIMS	Advanced Integrated MAD System
AIMS	Aircraft Identification Monitoring System (USA)
AIMS	Arctic Ice Monitoring System
AIMS	Attitude Indicator Measurement System
Aimval	Air Intercept Missile Evaluation (USA)
AINS	Airborne Inertial Navigation System
AIO	Action Information Organisation
AIP	Air Independent Propulsion
AIPS	Advanced Integrated Propulsion System (USA)
AIPT	Advanced Image Processing Terminal
AIR	Air Intercept Radar
AIR	Annual Infrastructure Report
AIR	Artilleriyskaya Instrumental'naya Razvedka. Artillery Instrument Reconnaissance (CIS)
Air Cdre	Air Commodore
AIRBALTAP	Allied Air Forces, Baltic Approaches (NATO)
AIRCOM	Air Command (Canada)
AIRNON	Allied Air Forces, Northern Norway (NATO)
AIRNORECHAN	Maritime Air Command, Nore Sub-Area (UK)
AIRNORLANT	Maritime Air Northern Sub-Area Command (UK)
AirOpsCen	Air Operations Centre (NATO)
AIRS	Advanced Inertial Reference System
AIRS	Advanced Infra-Red imaging Seeker
AIRS	Airborne Integrated Reconnaissance System
AIRSONOR	Allied Air Forces, Southern Norway (NATO)
AIRSOUTH	Allied Air Forces, Southern Europe (NATO)
AIRT	Air-to-air Interceptor Radar Training
AIRTO	Association of Independent Research & Technology Organizations
AIS	Advanced Instrumentation Subsystem
AISA	Aeronautica Industrial SA (Spain)
AISEC	Army Information Engineering Command (US Army)
AITAT	Armoured Infantry Training & Advisory Team
AIU	Airborne Installation Unit
AIU	Aircraft Interface Unit
AIU	Armament Interface Unit
AIU	Audio Interface Unit
AIWS	Advanced Interdiction Weapon System
AJ	Anti-Jamming

AJM	Anti-Jam Modem
AJT	Advanced Jet Trainer
AK	Avtomat Kalashnikov (CIS)
AK	Cargo ship (USA)
AKG	Akustische und Kino-gerate GmbH (Austria)
AKM	AK, Modified
AKR	Vehicle cargo ship (USA)
ALAAVS	Advanced Light Armored Amphibious Vehicle System (USA)
ALAD	Automatic Liquid Agent Detector (USA)
ALADI	Asociacion Latinoamericana de Integracion
ALAF	Asociacion Latinoamericana de Ferrocarriles
ALALC	Asociacion Latinoamericana de Libre Comercio
ALAMAR	Asociacion Latinoamericana de Aramadores
ALAMO	Affut Leger Anti-aerien Mobile. Launcher system (France)
ALARM	Air-Launched Anti-Radiation Missile
ALARMS	Airborne Laser Radar Mine Sensor
ALARR	Air-Launched, Air-Recoverable Rocket
ALB	Automatic Load-dependent Brake
ALBI	Affut Leger Bimunition. Lightweight twin-round Mistral missile launcher system (France)
ALBM	Air-Launched Ballistic Missile
ALC	Air Logistics Command (USA)
ALC	Automatic Level Control
ALCAM	Air-Launched Conventional Attack Missile (USA)
ALCC	Airborne Launch Control Centre
ALCM	Air-Launched Cruise Missile
ALCOA	Aluminum Co. of America
ALCOR	ARPA Lincoln Coherent Observable Radar (USA)
ALCS	Airborne Launch Control System
ALCV	Armoured Landmine Clearing Vehicle (Germany)
ALDP	Airborne Laser Designator Pod
ALE	Army Air Corps (Italy)
ALE	Automatic Link Establishment
ALERTDECL	Alert Stage / State Declaration
ALERTS	Airborne Laser Electronic warfare Receiver Training System (USA)
ALEX	Automatic Launching of Expendables
ALF	Atomic Line Filter
ALFENS	Automated Low Flying Entry & Notification System
ALFS	Airborne Low Frequency Sonar
ALGMS	Advanced Laser-Guided Missile System
ALGOL	Algorithmic Language
ALH	Active Laser Homing
ALH	Advanced Light Helicopter
ALI	Automatic Line Integration
ALLA	Allied Long Lines Agency
ALLD	Airborne Laser Locator Designator
ALLTV	All Light Level Television
ALM	Air Loadmaster (Royal Air Force)
ALMV	Air-Launched Miniature Vehicle
ALOC	Air Line of Communications

ALOFT	Airborne Light Optical Fibre Technology (USA)
ALOS	Airborne Light Observation System (South Africa)
ALOTS	Airborne Lightweight Optical Tracking System
ALPS	Accidental Launch Protection System
ALQA	Automatic Link Quality Analysis
ALR	Arbeitsgruppe fur Luft-und Raumfahrt (Switzerland)
ALRAAM	Advanced Long-Range Air-to-Air Missile
ALRAD	Airborne Laser Rangefinder And Designator
ALRI	Airborne Long-range Radar Input
ALS	Advanced Launch System (SDI project) (USA)
ALS	All-weather Landing System
ALS	American Laser Systems Inc.
ALS	Auxiliary Lighter (USA)
ALSS	Air-Launched Saturation System missile (USA)
ALT	Airborne Laser Tracker
ALT	Altimeter
ALT	Armoured Launching Turret
ALT	Attack Light Torpedo
ALTA	Advanced Lightweight Tactical Antenna
ALTAIR	ARPA Long-range Tracking And Instrumentation Radar (USA)
ALW	Air / Land Warfare
ALWT	Advanced Lightweight Torpedo
AM	Air Marshal
AM	Amplitude Modulation
AM	Anti-Material
AMA	Adaptive Multi-function Antenna
AMA	Antennenmastanlage. Telescopic antenna mast system (Germany)
AMAD	Airframe Mounted Accessory Drive
AMARV	Advanced Manoeuvring Re-entry Vehicle
AMAS	Area Maritima del Atlantico Sur (Argentina)
AMAS	Automated Manoeuvring Attack System
AMASE	Aircraft Maintenance And Support Equipment Exhibition & Conference
AMC	Advanced Micro-Electronics Converter
AMC	Air Materiel Command (Denmark)
AMC	Air Mobility Command (USA)
AMC	Army Materiel Command (USA)
AMC	The Armstrong Monitoring Corp. (Canada)
AMCA	Air Movement & Control Association (USA)
AMCCOM	Armament Munitions & Chemical Command (US Army)
AMCM	Advanced Mine Countermeasures
AMCS	Airborne Missile Control System
AMCS	Alert Monitoring System Computer
AMCSA	Army Materiel Command Support Activity Office (US Army)
AMDR	Automatic Missile Detection Radar
AMDS	Anti-Missile Discarding Sabot
AMDS	Automatic Manoeuvring Device System
AME	Amplitude Modulation Equivalent
AME	Angle Measuring Equipment
AME	Ateliers Maintenance Electronique (France)

AMECON	Australian Marine Engineering Consolidated Ltd
AMELEX	American Electronics Inc.
AMES	Advanced Multiple Environment Simulator
AMETS	Artillery Meteorological System (UK)
AMEWS	Automated Mobile EW System
AMF	ACE Mobile Force (NATO)
AMF	Amphibische Mehrzweek-Fahzeuge. Multi-purpose amphibious vehicle (Germany)
AMF-A	ACE Mobile Force – Air (NATO)
AMFCTS	Artillery & Mortar Fire Control Training Simulator
AMF-L	ACE Mobile Force – Land (NATO)
AMG	Antenna Mast Groups
AMG	Ateliers Mecaniques de Saint Gaudens (France)
AMGOT	Allied Military Government of Occupied Territories
AMHMS	Advanced Magnetic Helmet-Mounted Sight
AMHS	Automated Message Handling System
AMI	Aeronautica Militare Italiana
AMIDARS	Airborne Minefield Detection And Reconnaissance System
AMIDS	Airborne Minefield Detection System
AMIR	Anti-Missile Infra-Red
AMIS	Anti-Materiel Incendiary Submunition
AML	Automitrailleuse Leger. Light armoured car (France)
AMM	Anti-Missile Missile
AMOP	Amplitude Modulation On Pulse
AMOS	ARPA Maul Optical Station (USA)
AMP	Air Member for Personnel
AMP	Aviation Modernization Plan (US Army)
AMPM	Aircraft Missile Parts Mfg. Co. (USA)
AMPS	Automated Message Processing System (UK)
AMPSS	Advanced Manned Precision Strike System
AMR	Automitrailleuse de Reconnaissance (France)
AMRAAM	Advanced Medium Range Air-to-Air Missile
AMRAD	Armament / Munitions Requirements & Development Committee (USA)
AMRAD	Automatic Message Routeing And Distribution
AMRS	Advanced Maintenance Recorder System
AMS	Advanced Missile System (US Navy)
AMS	Aerospace Materials Specification (USA)
AMS	Air Maintenance Squadron (USA)
AMS	Air Management System
AMS	Airspace Management Systems SA/NV (Belgium)
AMS	American Management Systems Inc.
AMS	American Material Standard
AMS	Appareillages et Materiels de Servitudes SA (France)
AMS	Armoured Mortar System (UK)
AMS	Automatic Marking System
AMS	Automatic Meteorological System
AMS	Avionics Management System
AMSG	Allied Military Security General (NATO)
AMS-H	Advanced Missile System – Heavy

AMSO	Air Member for Supply & Organisation
AMSS	Advanced Multi-Sensor System
AMT	Active Memory Technology Inc. (USA)
AMTD	Adaptive Moving Target Detection
AMTI	Airborne Moving Target Indicator
AMU	Antenna Matching Unit
AMU	Auxiliary Memory Unit
AMV	Anti-Materiel et Vehicule. Anti-vehicle & equipment (France)
AMV	Armored Maintenance Vehicle
AMVET	Federal Division of Military Veterinary Service (Switzerland)
AMW	Advanced Microwave
AMX	Atelier de Construction d'Issy-les-Moulineaux (France)
AN	Air Force / Navy (USA)
ANA	Army, Navy, Air Force
ANC	African National Congress
ANCA	Allied Naval Communications Agency
ANDVT	Advanced Narrow-band Digital Voice Terminal (USA)
ANF	Atlantic Nuclear Force (UK)
ANG	Air National Guard (USA)
ANI	Automatic Number Identification
ANIE	Associazione Nazionale Industrie Elettrotecniche ed Elettroniche (Italy)
ANMI	Air Navigation Multiple Indicator
ANN	Automatic Neural Networks
ANO	Abu Nidal Organization
ANODE	Ambient sea Noise Directionality Estimator
ANP	Aircraft Nuclear Propulsion
ANP	Appareil Normal de Protection (France)
ANP	Armee National Populaire (Algeria)
ANRI	Automatic Net Radio Interfacing
ANS	Airborne Navigation System
ANS	American Nuclear Society
ANS	Anti-Navire Supersonique. Supersonic anti-ship missile (France)
ANS	Automatic Navigation System
ANSI	American National Standards Institute
ant	Antenna
ANT	Air Navigation Trainer
ANV	Advanced Naval Vehicle
ANVIS	Aviator's Night Vision Imaging System
ANZAAS	Australian & New Zealand Association for the Advancement of Science
ANZAC	Australia & New Zealand Corps
ANZUK	Australia, New Zealand & United Kingdom military alliance
ANZUS	Australia, New Zealand & USA security treaty
AO	Army Order
AO	Artillery Observation
AO	Fragmentation bomb (CIS)
AO	Oiler (USA)
AOA	Airborne Optical Adjunct
AOA	Amphibious Operating Area (USA)

AOA	Angle Of Attack
AOB	Air Order of Battle
AOB	Angle Of Bank
AOB	Arctic Ocean Buoy
AOB	Automatic Optical test-Bench
AOC	Adaptive Optical Camouflage
AOC	Air Officer Commanding
AOC	Air Operations Center (USA)
AOC	Association of Old Crows (USA)
AOC	Assumption Of Control (message)
AOC in C STC	Air Officer Commander-in-Chief Strike Command (NATO)
AOCINC	Air Officer Commanding in Chief
AOCM	Advanced Optical Countermeasures
AOCS	Attitude & Orbit Control System
AOCSTC	Air Officer Commander-in-Chief, Strike Command
AOD	Area Of Display
AOE	Fast Combat Support Ship (USA)
AOF (L)	Large fleet tanker
AOF (S)	Small fleet tanker
AOFR	Australian Optical Fibre Research Pty Ltd
AOG	Gasoline tanker (USA)
AOI	Arab Organization for Industrialization (Egypt)
AOI	Area Of Interest
AOMC	Aerojet Ordnance & Manufacturing Company (USA)
AO-N	Administrative Office, Navy (USA)
AOP	Airborne Observation Post
AOP	Area Of Projection
AOPF	Active Optical Proximity Fuze
AOPR	Area Of Prime Responsibility
AOPT	Advanced Optical Position Transducer
AOR	Replenishment Oiler (USA)
AOS	Add-On Stabilisation
AOS	Airborne Optical Sensor
AOS	Support Tanker
AOSA	Airborne Optical Sensor Adjunct
AOSNI	Air Officer Commanding, Scotland & Northern Ireland
AOSP	Advanced On-board Signal Processor
AOT	Transport oiler (USA)
AOTC	Associated Offices Technical Committee
AOU	Area Of Uncertainty
AOV	Artillery Observation Vehicle
AP	Air Publication
AP	Allied Publication
AP	Ammonium Perchlorate
AP	Anti-Personnel
AP	Armour-Piercing
AP	Transport Ship (USA)
AP2C	Appareil Portable pour la controle de la Contamination Chimique. Portable chemical contamination monitor (France)
APA	Adaptive Power Applique

APA	Advanced Pilot's Aid
APA	Amphibious transport
APACHE	Advanced Phased Array Chemical High Energy laser equipment
APACHE	Analysis of Pacific Area Communications for Hardening to EMP (USA)
APACHE	Arme Propulsee A Charge Ejectables. Submunitions dispenser system (France)
APAG	Atlantic Policy Advisory Group
APAM	Anti-Personnel, Anti-Material
APAR	Advanced Phased-Array Radar
APB	Analysing Printer Base
APB	Self-propelled barracks ship (USA)
APC	Aircraft Parts Corp. (USA)
APC	Armoured Personnel Carrier
APC	Armour-Piercing, Capped
APC	Automatic Phase Control
APCBC	Armour-Piercing, Capped, Ballistic Capped
APC-T	Armour-Piercing, Capped, Tracer
APCT-BF	Armour-Piercing, Capped, Tracer, Base Fuze
APCU	Alarm & Power Remote-Control Unit
APDS	Armour-Piercing, Discarding Sabot
APDS-T	Armour-Piercing, Discarding Sabot, Tracer
APE	Amphibisches Pionier-Erkundungsfahrzeug. Amphibious front line reconnaissance vehicle (Germany)
APE	Armour Piercing, Explosive
APE	Automated Production Equipment Corp. (USA)
APEI	Armour Piercing, High Explosive & Incendiary
APERS	Anti-Personnel
APERS-T	Anti-Personnel, Tracer
APES	Azimuth Position & Elevation System
APFC	Air-Portable Fuel Container
APFD	Autopilot Flight Director
APFIDS	Armour-Piercing, Fragmentation Incendiary, Discarding Sabot
APFSDS	Armour-Piercing, Fin-Stabilised, Discarding Sabot
APFSDS-T	Armour-Piercing, Fin-Stabilised, Discarding Sabot, Tracer
APFSDS(P)	Armour-Piercing, Fin-Stabilised, Discarding Sabot, (Practice)
APG	Aberdeen Proving Grounds
APGM	Autonomous Percision-Guided Munition (USA)
AP-HC	Armour-Piercing, Hard Core
APHC	Armour-Piercing, High Capacity
APHCI	Armour-Piercing, Hard Core, Incendiary
APHE	Armour-Piercing, High Explosive
APHEI	Armour-Piercing, High Explosive, Incendiary
APHE-SD	Armour-Piercing, High Explosive, Self-Destroying
APHIDS	Advanced Panoramic Helmet Interface Demonstrator System
API	Aero Products International SAS (Italy)
API	Aim Point Initiative
API	Air Photo Interpreter (UK)
API	Air Position Indicator
API	American Petroleum Institute
API	Application Programming Interface

API	Armour-Piercing, Incendiary
APIHC	Armour-Piercing, Incediary, Hard Core
APIT	Armour-Piercing, Incendiary, Tracer
APL	Barracks craft (USA)
APLS	Advanced Procurement & Logistics Systems Conference
APM	Anti-Personnel Mine
APNV	Advanced Products NV (Belgium)
APOBS	Anti-Personnel Obstacle Breaching System
APP	Application Portability Profile
APP	Armor Protection Program (USA)
APQ	Armaments Planning Questionnaire
APR	Air Photo Reader (UK)
APR	Auto Power Reserve
APRA	Alianza Popular Revolucionaria Americana (Peru)
APRE	Army Personnel Research Establishment
APS	Adaptive Processor System (sonar)
APS	Advanced Fighter Crew Protection System
APS	Aerial Post Squadron (USAF)
APS	Aircraft Position Sensor
APS	Aircraft Prepared for Service
APSE	Ada Programming Support Environment
APSE	Armour-Piercing, Secondary Effect
APSE-T	Armour-Piercing, Secondary Effect-Tracer
APSI	Aircraft Propulsion System Integration (USA)
APSOH	Advanced Panoramic Sonar – Hull-mounted
APSS	Atlas Port Surveillance System
APT	Armour-Piercing, Tracer
APTE	Abrams Power Train Evolution (USA)
APU	Armoured Patrol Unit
APU	Auxiliary Power Unit
APV	Armoured Patrol Vehicle
APV	Autopiloted Vehicle
APVO	Aviatsiya Proivo Vozdushnoi Oborny. Former Soviet Air Defence Aviation (CIS)
APWT	Annual Personal Weapon Test
AQAP	Allied Quality Assurance Publication (NATO)
AQD	Aeronautical Quality Assurance Directorate
AQF	Advanced Quickfix sensor
AQL	Acceptable Quality Level
AQL	Advanced Quickfix sensor
AR	Army Regulation
AR	Assault Rifle
AR	Repair Ship (USA)
ARA	Aerospace Research Associates Inc. (USA)
ARA	Armada Republica Argentina
ARABEL	Antenne Radar a Balayage Electronique. Phased array radar antenna (France)
Aracor	Advanced Research & Applications (USA)
ARAF	Afghan Republican Air Force
ARBAT	Application of Radar to Ballistic Acceptance Testing of ammunition

ARBS	Angle Rate Bombing Set
ARC	Allied Research Corp. (USA)
ARC	Ames Research Center (USA)
ARC	Area Reprogramming Capability
ARC	Armada Republica de Colombia. Colombian Navy
ARC	Atlantic Research Corp. (USA)
ARC	Cable Repairing ship (USA)
ARCADS	Armament Control & Delivery System
ARCE	Amphibious River Crossing Equipment
ARCI	Allied Research Corp. International (UK)
ARCM	Anti-Radiation Countermeasures
ARCO	Atlantic Richfield Co. (USA)
ARCS	Acquisition Radar & Control System
ARCS	Aerial Rocket Control System
ARD	All Round Defence (UK)
ARD	Anti-Radar Drone
ARD	Auxiliary Repair Dry dock (USA)
ARDCO	Advance Ratio Design Co. Inc. (USA)
ARDEC	Armament Research, Development & Engineering Center (US Army)
ARDM	Medium Auxiliary Repair Dry dock (USA)
ARDNOT	Automatic Day / Night Optical Tracker
ARDS	Airborne Radar Demonstrator System
ARE	Admiralty Research Establishment (UK)
ARE	Airborne Radar Extension
ARE	Applicazioni Radio Elettroniche SpA (Italy)
ARE	Atelier de Construction Roanne (France)
AREN	Army Radio Engineered Network (India)
ARES	Active Relief Sensor
ARES	Artillerie – Raketeneinsatzsystem. Rocket artillery employment system (Germany)
ARETS	Armour Remote Target System
ARF	Air Reserve Forces (USA)
ARF	Anti-Radar Futur (France)
ARFA	Allied Radio Frequency Agency
ARFA	Apparatus Respiratoire Filtrant des Armees. Military respirator (France)
ARGS	Advanced Resolution Gunnery Simulator
ARH	Active Radar Homing
ARH	Anti-Radar / Radiation Homing
ARI	Airpower Research Institute (USAF)
ARIC	Airborne Radio & Intercom Control system
ARINC	Aeronautical Radio Incorporated (USA)
ARJS	Airborne Radar Jamming System
AR-L	Airborne Reconnaissance – Low
ARL	Repair ship, small (USA)
ARM	Anti-Radar /-Radiation Missile
ARM	Artillery Re-arm Module
ARM	Availability, Reliability, Maintainability
ARMAD	Armoured & Mechanised Unit Air Defence
ARMAT	Anti-Radar missile Matra (France)

ARM-D	Anti-Radiation Missile Decoy
ARMS	Advanced Radar Missile Scoring System
ARMS	Advanced Remote Minehunting System
ARMS	Armoured Resupply Maintenance System
ARMS	Atlantic Research Marketing Systems Inc. (USA)
ARMSCOR	Armament Manufacturing Corporation (South Africa)
ARMSCOR	Arms Corporation of the Philippines
ARMVAL	Anti-Armor Vehicle Evaluation (USA)
ARNG	Army National Guard (USA)
ARO	Army Research Office (USA)
ARO	Auxiliary Readout
ARP	Aerial Rotation Period
ARP	Air Raid Precautions
ARP	Alternative Release Procedure
ARP	Anti-Radiation Projectile
ARP	Attack Reference Point
ARPA	Advanced Research Projects Agency (USA)
ARPANET	Advanced Research Projects Agency Network (USA)
ARPS	Acoustic Range Prediction System
ARPS	Advanced Radar Processing Subsystem
ARPTT	Air-Refuelling Part-Task Trainer (USAF)
ARQ	Automatic Repeat Request
ARRADCOM	Armament Research & Development Command
ARRC	Allied Command Europe Rapid Reaction Corps (NATO)
ARRV	Armoured Repair & Recovery Vehicle
ARS	Advanced Rocket System
ARS	Aerial Radiac System
ARS	Air Reinforcement Squadron
ARS	Air Rescue Service (USAF)
ARS	Attack Radar Set (USA)
ARS	Automatic Reporting System
ARS	Salvage Ship (USA)
ARSOM	Artilleriyskaya Radiolokatsionnaya Stantsiya Obnaruzheniya Minometov. Mortar location radar (CIS)
ARSR	Air Route Surveillance Radar
ARSV	Armoured Reconnaissance Scout Vehicle (USA)
ART	Air-Refuelling Transport
ARTADS	Army Tactical Data System (USA)
ARTBASS	Army Training Battle Simulation System
ARTCC	Air Route Traffic Control Centre
ARTHUR	Artillery Hunting Radar (Sweden)
ARTI	Advanced Rotocraft Technology Integration
ARTI	Aeronautical Research & Test Institute (Czechoslovakia)
ARTIST	Advanced Radar Techniques for Improved Surveillance & Tracking
ARTS	All-Round Thermal Surveillance
ARV	Armoured Recovery Vehicle
ARWAR	Advanced Radar Warning Receiver
ARWS	Advanced Radar Warning System
AR/AAV	Armored Reconnaissance / Airborne Assault Vehicle (USA)
AS	Air-to-Surface

AS	Ammunition Storage
AS	Submarine Tender (USA)
ASA	Advanced Security Agency
ASA	Advanced System Architecture
ASA	Advanced System Architectures Ltd (UK)
ASAAC	Allied Standard Avionics Architecture Council
ASAC	Airborne Surveillance, Airborne Control
ASALA	Armenian Secret Army for the Liberation of Armenia
ASALM	Advanced Strategic Air-Launched Missile
ASAM	Advanced Surface-to-Air Missile
ASAP	Airborne Shared Aperture Program (USA)
ASARC	Army Systems Acquisition Review Council
ASARG	Autonomous Synthetic Aperture Radar Guidance
ASARS	Advanced Synthetic Aperture Radar System
ASARS	Airborne Search & Rescue System
ASAS	All Source Analysis System (USA)
ASASCO	Allied-Signal Aerospace Service Corp. (France)
ASAT	Advanced Subsonic Aerial Target
ASAT	Anti-Satellite (weapon or system)
ASAT	Avionics Situation Awareness Trainer
ASBM	Air-to-Surface Ballistic Missile
ASC	Accredited Standards Committee
ASCA	Army Scaling & Codification Authority
ASCB	Aircraft Standard Communication Bus
ASCC	Air Standardization Co-ordination Committee
ASCII	American Standard Code for Information Interchange
ASCL	Advanced Sonobuoy Communications Link (USA)
ASCM	Advanced Spaceborne Computer Module
ASCM	Anti-Ship Cruise Missile
ASCON	Automatic Switched Communications Network (USA)
ASCOT	Airspace Control & Operations Training simulator (USA)
ASCU	Armament Station Control Unit
ASC/PAM	Advanced System Control / Processor And Memory
ASD	Advanced System Development
ASDC	Alternative Space Defence Centre
ASDC	Armament Signal Data Converter
ASDC	Army Strategic Defense Command (USA)
ASDI	Analogue Simple Data Interface set
ASDIC	Allied / Anti-Submarine Detection & Investigation Committee
ASDL	Airborne Self-Defence Laser
ASDP	Advanced Seeker Development Program (USA)
ASE	Aero Systems Engineering Inc. (USA)
ASE	Aircraft Survivability Equipment
ASE	Automatic Stabilisation Equipment
ASEAN	Association of Southeast Asian Nations
ASEM	Anti-Ship Euromissile consortium
ASEMA	Advanced Special Electronics Mission Aircraft (US Army)
ASES	Active Sonar for Entrance Surveillance
ASET	Automatic Scoring Electronic Target
ASETS	Airborne Seeker Evaluation Test Set

ASF	Army Standard Family (shelters)
ASG	Airborne System, Gun (USA)
ASG	Assistant Secretary General (NATO)
ASG	Ausbildungsgerate Schuss / Getroffen (Germany)
ASGC	Airborne Surveillance, Ground Control
ASGW	Air-to-Surface Guided Weapon
ASH	Advanced Seeker Homing
ASH	Advanced Support Helicopter (USA)
ASI	Airport Systems International Inc. (USA)
ASI	Airspeed Indicator
ASI	Australian Shipbuilding Industries (WA) Pty Ltd
ASI	Aviation Services International (Benelux) BV (Netherlands)
ASI	Avionics Specialist Inc. (USA)
ASI	Italian Space Agency
ASIC	Application Specific Integrated Circuit
ASIL	American Society of International Law
ASIP	Aircraft Structural Integrity Program (USA)
ASIR	Airspeed Indicator Reading
ASJ	Automatic Search Jammer
ASL	Acting Sub-Lieutenant
ASL	Applied Science Laboratories (USA)
ASL	Authorised Stockage List
ASL	Aviation Spares Ltd (UK)
ASLO	Accident Site Liaison Officer
ASLP	Air-Sol Longue Portee (France)
ASLR	Air / Surface Laser Ranger
ASM	Advanced Sea Mine
ASM	Air-to-Surface Missile
ASM	Armored Systems Modernization programme (USA)
ASMA	Air Staff Management Aid. Now an RAF owned CCIS
ASMAR	Astilleros y Maestranzas de la Armada (Chile)
ASMD	Anti-Ship Missile Defence
ASME	American Society of Mechanical Engineers
ASMI	Airport Surface Movement Indicator (radar)
ASMP	Air-Sol Moyenne Portee. Air-to-surface medium- range missile (France)
ASMS	Advanced Surface Machinery System (US Navy)
ASMS	Advanced Surface Missile System
ASN-1	Abstract Syntax Notation One
ASNE	American Society of Naval Engineers
ASNI	Ambient Sea Noise Indication
ASOC	Air Support Operations Centre
ASOM	Autonomous Stand-Off Missile (Sweden)
ASP	Advanced Signal Processor
ASP	Airbase Survivability Program (USA)
ASPJ	Airborne Self-Protection Jammer
ASPR	Armed Services Procurement Regulation
ASPRO	Advanced Signal Processor
ASQC	American Society of Quality Control
ASR	Acoustic Self-Ranging system

ASR	Air Staff Requirement
ASR	Approach Surveillance Radar
ASR	Automatic Send / Receive
ASR	Submarine Rescue ship (USA)
ASRAAM	Advanced Short-Range Air-to-Air Missile
ASRAO	Advanced Systems Research & Analysis Office (USA)
ASRAP	Acoustic Sensor Range And Prediction
ASRM	Advanced Solid Rocket Motor (NASA)
ASROC	Anti-Submarine Rocket
ASRP	Artillery Systems Research Programme
ASRT	Air Support Radar Team
ASS	Ammunition Storage Sites
ASS	Anti-Shelter Submunition
ASS	Aviation Support Ship
ASSATS	Advanced Ship-to-Shore Automatic Telegraph System (UK)
ASSESSREP	Assessment Report
ASSG	Acoustic Sensor Signal Generator
ASSM	Anti-Ship Supersonic Missile
ASST	Anti-Ship Surveillance & Targeting
ASSTASS	Advanced Surface Ship Towed Array Surveillance System
ASSW	Anti-Surface Ship Warfare
AST	Advanced Simulation Technology
AST	Advanced Systems Technology Inc. (USA)
AST	Air Service Training Ltd (UK)
AST	Air Staff Target (UK)
AST	Atlantic Standard Time
AST	Avionics Systems Trainers
ASTA	AeroSpace Technologies of Australia Pty Ltd
ASTABS	Automated Status Board
ASTAR	Advanced Series of Tactical Air Defence Radars
ASTAR	Airborne Search Target Attack Radar
ASTARTE	Avion-Station-Relais de Transmissions Exceptionelles. Airborne communications relay system (France)
ASTAT	Assessment of Stationery Target Acquisition Techniques programme (USA)
ASTB	American Standards Testing Bureau Inc.
ASTI	Aviation Support Technology Int'l Inc. (USA)
ASTM	American Society for Testing & Materials
ASTOR	Airborne Stand-Off Radar
ASTRAL	Air Surveillance & Targeting Radar, L-band
ASTROS	Artillery Saturation Rocket System
ASTT	Action Speed Tactical Trainer
ASTT	Attack Submarine Team Trainer
ASU	Acoustic Simulation Unit
ASU	Active Sonar (Germany)
ASU	Automatic Switching Unit
ASuW	Anti-Surface vessel Warfare
ASUW	Anti-Surface Unit Warfare
ASV	Ammunition Supply Vehicle
ASV	Anti-Ship Variant radar

ASV	Anti-Surface Vessel
ASVEH	Air-Surveillance Vehicle
ASVU	Army Security Vetting Unit (UK)
ASVW	Anti-Surface Vessel Warfare
ASW	Anti-Submarine Warfare
ASWAC	Anti-Submarine Warfare Analysis Centre
ASWACS	Airborne Surveillance And Warning Control System (India)
ASWAS	Anti-Submarine Warfare Area System
ASWDS	Anti-Submarine Warfare Data System
ASWE	Admiralty Surface Weapons Establishment (UK)
ASWHQ	Alternative Subordinate War Headquarters
ASWT	Anti-Submarine Warfare Trainer
ASW/SOW	Anti-Submarine Warfare / Stand-Off Weapon
ASXV	Air-launched Expendable Sound Velocimeter
ASZ	Air Surface Zone
AS&E	American Science & Engineering Inc.
AS/GC	Air Surveillance & Gun Control console
AT	Advanced Trainer (USA)
AT	Anti-Tank
ATA	Actual Time of Arrival
ATA	Advanced Tactical Aircraft
ATA	Advanced Test Accelerator
ATA	Aerospace Technologies of Australia
ATA	Air Transport Association of America
ATA	Air Transport Auxiliary
ATA	Atlantic Treaty Association
ATA	Auxiliary Ocean Tug (USA)
ATAAC	Anti-Torpedo Air-launched Acoustic Countermeasures
ATAC	Air-Transportable Acoustic Communications buoy
ATAC	All-Terrain All Climate
ATAC	All-Terrain Amphibious Carrier
ATAC	Applied Technology Advanced Computer
ATACC	Advanced Tactical Command Central
ATACDIG	Army Tactical Digital Image Generator
ATACMS	Army Tactical Communications Management System
ATACMS	Army Tactical Missile System (USA)
ATACO	Air Tactical Control Operator
ATACS	Advanced Tank Cannon System (USA)
ATACS	Advanced Target Acquisition Counterfire System (US Army)
ATACS	Army Tactical Communications System (USA)
ATAF	Allied Tactical Air Force (NATO)
ATAFCS	Airborne Target Acquisition & Fire Control System
ATAGS	Advanced Technology Anti-G Suit (USA)
ATAL	Appareillage de Télévision sur Aeronef Leger (France)
ATAM	Air-To-Air Mistral missile (France)
ATAR	Air-To-Air Recognition device
ATAR	Air-To-Air Recovery
ATARS	Advanced Tactical Air Reconnaissance System
ATAS	Advanced Target Acquisition Sensor
ATAS	Air-To-Air Stinger missile

ATATS	Automatic Target Acquisition & Tracking System
ATB	Advanced Technology Bomber
ATBM	Anti-Tactical Ballistic Missile
ATC	Acoustic Torpedo Countermeasures
ATC	Air Traffic Control
ATC	Air Training Command (USAF)
ATC	Air Training Corps (UK)
ATC	Automatic Threat Countering
ATC	Automatic Tuning Control
ATC	Mini-Armored Troop Carrier (USA)
ATCA	Allied Tactical Communications Agency (NATO)
ATCAP	Army Telecommunications Automation Program (USA)
ATCCS	Army Tactical Command & Control System (USA)
ATCRBS	Air Traffic Control Radar Beacon System
ATCS	Advanced Tank Cannon System
ATCS	Air Traffic Control Subsystem
ATD	Actual Time of Departure
ATD	Advanced Technology Demonstration
ATDL	Army Tactical Data Link
ATDS	Airborne Tactical Data System (USA)
ATDSIA	Allied Tactical Data Systems Interoperability Agency
ATDU	Armoured Trials & Development Unit
ATE	Automatic Test Equipment
ATEMS	Advanced Threat Emitter Simulator
ATER	Advanced Triple Ejector Rack
ATESS	Advanced Tactics & Engagement Simulation System
ATET	Azienda Torinese Elettronica Telecomunicazioni (Italy)
ATF	Advanced Tactical Fighter
ATF	Air Transport Force
ATF	Fleet ocean tug (USA)
ATFCS	Advanced Tank Fire Control System
ATG	Anti-Tank Gun
ATG	Automatic Test Guide
ATGM	Anti-Tank Guided Missile
ATGW	Anti-Tank Guided Weapon
ATH	Automatic Target Handover
ATH	Autonomous Terminal Homing
ATHS	Automatic Target Handover System
ATI	Alliance Technique Industrielle (France)
ATILA	Automatisation du Tir de l'Artillerie. Artillery automation (France)
ATIM	Advanced Technology Insertion Module
ATIRCM	Advanced Threat Infra-Red Countermeasures System (USA)
ATIS	Anti-Tank Influence Sensor
ATIS	Automatic Terminal Information Service
ATL	Aero Tec Laboratories Inc. (USA)
ATLANTIC	Airborne Targeting Low Altitude Navigation Thermal Imaging & Cueing
ATLAS	Abbreviated Test Language for All Systems
ATLAS	Advanced Technology LADAR System (USA)
ATLAS	Azimuth Target Intelligence & Acquisition System (Israel)

ATLIS	Automatic Tracking Laser Illumination System
ATLV	Artillery Target Location Vehicle
ATM	Abri Technique Mobile. ISO Shelter (France)
ATM	Anti-Tactical Missile
ATM	Anti-Tank Mine
ATM	Asynchronous Transfer Mode
ATMDS	Anti-Tank Mine-Dispensing System
Atmos	Ammunition, Toxic Material Open Space
ATMP	All-Terrain Mobile Platform
ATMS	ASW Track Management System
ATO	Ammunition Technical Officer (UK)
ATOC	Allied Tactical Operations Centre (NATO)
ATOPS	Advanced Transport Operations System
ATP	Acquisition, Tracking & Pointing
ATP	Air Turbine Pump
ATP	Allied Tactical Publication (NATO)
ATP	Anti-Tactical Patriot
ATP	Attack Plot
ATP	Australian Technical Publications
ATPFC	Acquisition, Tracking & Pointing & Fire Control
ATR	Advanced Tactical Radar programme
ATR	Air Transportable Racking. ARINC standard box sizes
ATR	Automatic Target Recognition
ATR	Automotive Test Rig
ATRAN	Automatic Terrain Radar And Navigator
ATRJ	Advanced Threat Radar Jammer
ATS	Accelerator Test Stand
ATS	Acoustic Tracking System
ATS	Advanced Tracking System
ATS	Advanced Training Systems Inc. (USA)
ATS	Agile Target System
ATS	Air Traffic Services
ATS	Air Traffic System
ATS	Air Transport Support
ATS	Aircrew Training System (USA)
ATS	Applied Technology Services (Israel)
ATS	Atelier de Construction de Tarbes (France)
ATS	Auxiliary Territorial Service. Superseded by WRACS
ATS	Aviation Training Ship (Royal Navy)
ATS	Salvage & rescue ship (USA)
ATSF	Abri Technique Semi Fixe. TEMPEST-hardened shelter (France)
AT-ST	Artillery survey vehicle (CIS)
ATSU	Accelerator Test Stand Upgrade
Att	Attachment (UK)
ATT	Artillery Tactical Terminals
ATT	Automatic Target Tracking
ATTD	Advanced Technology Test Demonstrator
ATTS	Air-Transportable Towed System
ATTU	Atlantic-to-the-Urals (NATO)
ATTV	All Terrain Tow Vehicle

ATU	Aircraft Target Unit
ATU	Antenna Tuning Unit
ATV	All-Terrain Vehicle
ATVS	Advanced TV Seeker
ATW	Air Transport Wing (Japan)
AUG	Armee Universal Gewehr. Army Universal Gun (Austria)
AUM	Air-to-Underwater Missile (USA)
AURA	Autonomous Un-manned Reconnaissance Aircraft
AURORA	Automatic Recovery Of Remotely piloted Aircraft
AUS	Assistant Under Secretary of State
AUSA	Association of the US Army
AUSTACCS	Australian Automatic Command & Control System
AUTODIN	Automatic Digital Network
AUTOKO	Automatische Korpsstamunetz. Automated Corps communication network (Germany)
AUTOSEVOCOM	Automatic Secure Voice Communications system (USA)
AUTOVON	Automatic Voice Network (USA)
AUV	Autonomous Underwater Vehicle
AUVS	Association for Unmanned Vehicle Systems (USA)
AUW	All-Up Weight
AUWE	Admiralty Underwater Warfare Establishment (UK)
Aux	Auxiliary
AV	Anti-Vehicle
AVA	Adaptive Vehicular Antenna
AVAD	Automatic Voice Alerting System
AVADS	Autotrack Vulcan Air Defense System (USA)
AVD	Atmospheric Vehicle Detection
AVDS	Armoured Vehicle Driving Simulators
AVE	Airborne Vehicle Equipment
Avgas	Aviation Gasoline
AVGP	Armoured Vehicle, General Purpose
AVH	Armoured Vehicle, Heavy
AVIP	Avionics Integrity Program (USAF)
AVL	Armoured Vehicle, Light
AVL	Automated Vehicle Location
AVLB	Armoured Vehicle-Launched Bridge
AVLF	Airborne Very Low Frequency
AVM	Air Vice-Marshall
AVM	Airborne Vapour Monitor
AVM	Armoured Vehicle, Medium
AVM	Guided missile ship (USA)
AVMRL	Armoured Vehicle Multiple Rocket Launcher
AVP	Advanced Video Processor
AVR	Armoured Vehicle, Reconnaissance
AVR	Army Volunteer Reserves
AVRE	Armoured Vehicle, Royal Engineers
AVRS	Airborne Video Recording System
AVS	Advanced Vertical Strike
AVS	Avimo Singapore Ltd
AVSCOM	Aviation Systems Command (US Army)

AVT	Analogue Voice Terminal
AVT	Automatic Video Tracker
AVT	Auxiliary Aircraft Landing Training Ship (USA)
Avtur	Aviation Turbine Kerosene
AW	All-Weather
AW	Amphibious Warfare ship
AW	Automatic Weapon
AWACS	Airborne Warning & Control System (USA)
AWADS	Adverse Weather Aerial Delivery System
AWARDS	Advanced Wide-Angle Reflective Display System
AWARE	Advanced Warning of Active Radar Emissions
AWCCV	Advanced Weapons Carriage Configured Vehicle
AWCLS	All-Weather Carrier Landing System
AWCS	Automatic Weapons Control System
AWDATS	Automatic Weapon Data Transmission System (UK)
AWDREY	Atomic Weapon Detection, Recognition & Estimation of Yield
AWE	Atomic Weapons Establishment (UK)
AWES	Area Weapons Effect Simulator
AWESS	Automatic Weapons Effects Signature Simulator
AWG	Air Weapons Group
AWHQ	Alternative War Headquarters (NATO)
AWITEL	Albis Wire Telephone (Switzerland)
AWLS	All-Weather Landing System
AWNIS	Allied Wartime Navigational Information System (NATO)
AWOL	Absent Without Official Leave
AWRE	Atomic Weapons Research Establishment
AWS	Air Weather Service (USA)
AWS	American Welding Society
AWSACS	All-Weather Stand-off Aircraft / Attack Control System (US Navy)
AWSAS	All-Weather Stand-off Attack System
AWTSS	All-Weather Tactical Strike System
AWX	All-Weather Interceptor
AX	Attack Experimental
AXBT	Air-launched Expendable Bathythermograph
AXCON	Axis Controller
A&AEE	Aeroplane & Armament Experimental Establishment (UK)
A&T	Analysis & Technology Inc. (USA)
A/C	Aircraft
A/D	Air Defence
A/E	Architecture / Engineering
a/f	Airfield
a/f	Airframe
A/G	Air-to-Ground
A/S	Anti-Submarine

B

BA	Bertan High Voltage (USA)
BA	British Association (for the Advancement of Science)
BAA	Broad Agency Announcement
BABS	Beam / Blind Approach Beacon System
BADGE	Base Air Defence Ground Equipment
BAe	British Aerospace plc
BAEE	British Army Equipment Exhibition
BAeSL	British Aerospace Simulation Ltd
BAFB	Bolling Air Force Base (USAF)
BAI	International Board of Auditors (NATO)
BAKWVT	Bundesakademie fur Wehrverwaltung und Wehrtechnik (Germany)
BAL	Bristol Aerospace Ltd (Canada)
BAL	Burdeshaw Associates Ltd (USA)
BALMET	Ballistic Trajectory calculators allowing for Meteorological conditions
BALO	Brigade Air Liaison Officer (UK)
BALTAP	Allied Forces, Baltic Approaches (NATO)
BALUN	Balancing Unit (antenna)
BAMF	Swiss maintenance organisation
BAOR	British Army of the Rhine
BAOS	Basic Artillery Observer Sub-system
BAP	Bombes Anti-Personnel. Anti-personnel bomb (France)
BAP	Bridge Adapter Pallet
BAP	Buque Aramada Peruana. Peruvian Navy
BARB	Boosted Anti-Radar Bomb (South Africa)
BARC	Beach Amphibious Resupply Cargo
BARV	Beach Armoured Recovery Vehicle
BASE	British Aerospace Systems & Equipment Ltd
BASEDAM	Base Damage Report
BASEEFA	British Approval Service for Electrical Equipment in Flammable Atmospheres
BASEGRAMME	Message sent to a unit's home base for onward transmission
BASG	Ball Corp. Aerospace Systems Group (USA)
BASI	Beech Aerospace Services Inc. (USA)
BASIC	British American Security Information Council (UK)
BASO	Brigade Air Support Officer
BAS(W)	British Army Staff (Washington)
bat	Battalion
BAT	Basic Aviation Trainer
BAT	Bavaria Avionik Technologie GmbH (Germany)
BAT	Bombes d'Appui Tactique. Tactical support bomb (France)
BAT	Brilliant Anti-Tank guided anti-armour submunition (USA)
BATCO	Battle Code (UK)
BATES	Battlefield Artillery Target Engagement System (UK)
BATR	Bullets At Target Range
BATS	Ballistic Aerial Target System
BATS	Battle Area Tactical Scenario
BATS	Battle Area Tactical Simulation

BB	Base Bleed
BB	Base Burn
BB	Battery Block
BB	Battleship (USA)
BB	Blowback
BBA	British Board of Agreement
BBC	Broad-Band Chaff
BBC	Bromobenzylcyanide. Chemical warfare tear agent
BBC	Built-in Ballistic Computer
BBE	Bridge Boat Erection
BBGT	Brigade & Battle Group Trainer
BBN	Bolt Baraneck & Newman
BBS	Brigade & Battalion Simulation
BBSAG	Battleship Surface Action Group
BBU	Base Bleed Unit
BC	Battery Commander
BC	Bayonet Cap
BC	Bomber Command
BCA	Buoyant Cable Antenna
BCAR	British Civil Airworthiness Requirement
BCAU	Backup Control & Audio Unit
BCB	Broadband Communications Bus
BCC	Battery Control Centre
BCD	Binary Coded Decimal
BCD	Bulk Chaff Dispenser
BCE	Battlefield Control Element
BCF	Briefing Co-ordination Facility
BCH	Boots Combat High (UK)
BCM	Basic Combat Manoeuvring
BCN	Beacon radar mode
BCP	Battery Control Post
BCR	Bomblets Cargo Round
BCS	Ballistic Computer System
BCS	Battery Computer System
BCS	British Calibration Service
BCS	British Computer Society
BCS	Broadcast Control Segment
BCT	Battlefield Command Terminal
BCT	Battlefield Command Trainer
BCU	Battery Computer Unit
BCU	Battery Coolant Unit
BCU	Bird Control Unit
BCW	Binary Chemical Warhead
BC/WC	Ballistic Computer / Weapon Controller
BD	Base Detonating
BDA	Battle Damage Assessment
BDA	Bomb Damage Assessment
Bde	Brigade
BDF	Belize Defence Force
BDI	Bearing Distance Indicator

BDI	Blast Deflectors Inc. (USA)
BDIG	Belgian Defence Industry Group
BDL	Bistable Diode Laser
BDLI	Bundesverband der Deutschen Luftfahrt, Raumfahrt und Ausrustungsindustrie e.V. (Germany)
BDLS	British Defence Liaison Staff
BDM	Bunker Defeat Munition (US Army)
Bdr	Bombardier
BDR	Base Drag Reduction
BDR	Battle Damage Repair
BDS	British Defence Staff
BE	Base Ejection
BEAB	British Electrotechnical Approvals Board
BEAMA	British Electrotechnical & Allied Manufacturers Association
BEAR	Beam Experiment Abroad Rocket
BEEO	Battlefield Electromagnetic Environments Office (USA)
BEES	Basic ECM Environment Simulator
BEF	British Expeditionary Force
BEL	Bharat Electronics Ltd (India)
BEM	Brush Electrical Machines Ltd (UK)
BEMA	British Equipment Manufacturers Association
BENECHAN	BENELUX-Sub Area Channel (NATO)
BENELUX	Belgium, Netherlands & Luxembourg
BEP	Burnley Engineering Products Ltd (UK)
BER	Beyond Economic Repair (UK)
BER	BIT Error Rate
BES	Busque de Efecto Superfice. Surface effect ship (Spain)
BET	Best Estimated Trajectory
BETA	Battlefield Exploitation & Target Acquisition
BEU	Bulk Encryption Unit
BF	Base Fuzed
BFD	Bearing / Frequency Display
BFDAS	Basic Flight Data Acquisition System
BFFI	British Forces, Falkland Islands
BFG	British Forces, Germany
BFHK	British Forces, Hong Kong
BFO	Beat Frequency Oscillation
BFPO	British Forces Post Office
BFPQ	Block Floating Point Quantization
BFT	Basic Fitness Test / Training (UK)
BFTA	Bulk Fuel Tank Assembly
BFVS	Bradley Fighting Vehicle System
BG	Bomb Group (USA)
BGT	Bodenseewerk Geratetechnik GmbH (Germany)
BHP	The Broken Hill Proprietary Co. Ltd (Australia)
BHT-E	Bell Helicopter Textron – Europe (UK)
BIC	Barfield Corp. (USA)
BICES	Battlefield Information Collection & Exploitation System (USA)
BICS	Battlefield Information Control System
BID	British Industrial Development

BIDS Battlefield Information Distribution System
BIFF Battlefield Identification Friend or Foe
BILL Bofors, Infantry, Light & Lethal
BIM Ballistic Intercept Missile
BIP Blown In Place (minefields)
BIR Biennial Infrastructure Review
BIRADS Bistatic Receiver And Display System
BIRI Batallones de Infanteria de Reaccion Immediata. Rapid reaction infantry battalion (El Salvador)
BIRIS Battlefield Inoculation Remote Initiation System
BISPA British Independent Steel Products Association
BIST Built-In Self Test
BIT Built-In Test
BITE Built-In Test Equipment
BJA Baseline Jamming Assets
BKEP Boosted Kinetic Energy Penetrator
BKT Bavaria Ketronic Gesellschaft fur Flugtechnische Gerate mbH (Germany)
BL Blank
BL Breech-Loading
BLC Bureau Interforces de'Codification (Belgium)
BLISK Blade plus Disk fabricated in one piece (turbine)
BLITS Beta Lighted Infantry Telescope System
BLR Beyond Local Repair (UK)
BLR Blindado Ligero de Ruedas (Spain)
BL-T Blank Tracer
BLU Bomb Live Unit
BM Ballistic Missile
BM Battle Management
BM Brigade Major
BMARC British Manufacture & Research Co.
BMC3 Battle Management Command Control & Communications
BMCC British Metal Castings Council
BMD Ballistic Missile Defence
BMDSCOM Ballistic Missile Defence Systems Command
BME Basic Mass Empty
BMEC British Marine Equipment Council
BMEWS Ballistic Missile Early Warning System (USA)
BMF Belgian Mechanical Fabrication
BMFT Bundesministerium fur Forschung und Technologie (Germany)
BMI Battelle Memorial Institute (USA)
BMMG British Microcomputer Manufacturers Group
BMO Ballistic Missile Organization (USA)
BMP Boyevaya Mashina Pekhoty. Former Russian APC (CIS)
BMR Blindado Medio de Ruedas (Spain)
BMS Battalion Mortar System
BMS Battlefield Management Systems
BMT British Maritime Technology Ltd
BMTR Basic Marksmanship Training Range
BMTS Ballistic Missile Target System

BMVg	Bundesministerium der Verteidigung (Germany)
Bn	Battalion
BN	Bangladesh Navy
BNEA	British Naval Equipment Association
BNFL	British Nuclear Fuels Ltd
BNFMF	British Non-Ferrous Metals Federation
BNRID	Basic Net Radio Interface Device
BNS	BILL Night Sight
BNSC	British National Space Centre
BOA	Basic Ordering Agreement
BOA	Broad Ocean Area
BOAMP	Bulletin Officiel des Annonces des Marches Publics. Government contracts bulletin (France)
BOBS	Beacon-Only Bombing System
BOC	Base Operating Centre
BOCV	Battery Operations Centre Vehicle
BOFI	Bofors Optronic Fire control Instrument
BOL	Bearing Only Launch
BOLD	Battlefield Operations Laser Designator
BOP	Balance of Payments
BOSS	Battalion Operated Surveillance System (USA)
BOSS	Bureau of State Security (South Africa)
BOTB	British Overseas Trade Board
BOW	Basic Operating Weight
BP	Black Powder
BPA	British Ports Association
BPDMS	Base Point Defence Missile System
BPE	Bomber Penetration Evaluation
BPI	B&P Instrumentation Ltd (UK)
BPR	Bypass Ratio
BPSK	Bi-Phase Shift Keying
BR	Book of Reference. (Royal Navy)
BRA	Bouganville Revolutionary Army (Papua New Guinea)
BRAB	Armour piercing bomb (CIS)
BRAVE	Boeing Robotic Air Vehicle (USA)
BRDEC	Belvoir Research, Development & Engineering Center (USA)
BRE	Battlefield Recovery & Evacuation
BRE	Building Research Establishment
BREM	Bronirovannaya Remontno-evakuatsionnaya. Armoured recovery & repair vehicle (CIS)
BREMA	British Radio & Electronic Equipment Manufacturers Association
Brig	Brigadier
Brig-Gen	Brigadier-General
BRITE	Basic Research in Industrial Technologies for Europe
BRITMISS	British Maritime Intelligence Support System
BRM	Basic Reference Model
BRP	Braked Retarded Parachute
BRPS	Super-Braked Retarded Parachute
BRSL	Bomb Release Safety Lock
BRU	Bomb Release Unit

BRW	Brake Release Weight
BS	British Standard
BSAF	Bosnian Serb Air Force
BSC	Belgian Shipbuilders Corporation
BSI	British Standards Institute
BSIA	British Security Industry Association (UK)
BSIN	Bus System Inferface Unit
BSIU	Bus Subsystem Interface Unit for cockpit management system
BSM	Barrett Soft Mount. Rifle mount
BSP	Barra Side Processor
BSP	Barrel Stave Projector. Acoustic projector
BST	British Summer / Standard Time
BSTS	Boost Surveillance & Tracking System
BSW	British Standard Whitworth screw thread
BT	Bathythermograph
BT	Bullet-Through
BT	Burn Time
BTA	Best Technical Approach
BTF	Baseline Test Facility
BTG	British Technology Group
BTH	Beyond The Horizon
BTI	Balanced Technology Initiative
BTO	Barbed Tape Obstacle
BTP	Bureau Trilateral de Programme
BTT	Basic Training Target
BTU	Bullet Trap Universal. MECAR patented (Belgium)
BUCS	Back-Up Control System
BUDFIN	Budget & Finance Division, SHAPE (NATO)
BUIC	Backup Intercept Control (for air defence)
BUS	Break-Up System
BV	Boost Vehicle
B-VIT	Base station Video Imaging Terminal
BVQI	Bureau Veritas Quality International Ltd
BVR	Beyond Visual Range
BW	Bacteriological Weapons
BW	Bandwidth
BW	Biological Warfare
BW	Bomb Wing (USA)
BWB	Bundesamt fur Wehrtechnik und Beschaffung (Germany)
BWC	Biological (Bacteriological) & Toxic Weapons Convention, 1972
BWVA	Bundeswehrverwaltungsamt (Germany)
BY	Bathythermograph
BZ	Beaten Zone (UK)
BZ	Benactyzine. Chemical warfare incapacitating agent
BZS	Bundesamtes fur Zivilschutz (Germany)
B&C	Biological / Bacteriological & Chemical weapons
B&GS	Bombing & Gunnery School
B/L	Bill of Loading

C

C2	Command & Control
C2A	Command, Control & Alerting. Airspace management network (France)
C2IPS	Command & Control Information Processing System
C2P	Command & Control Processor
C2S	Command, Control & Status
C3	Command, Control & Communications
C3CM	Command, Control & Communications Countermeasures
C3I	Command, Control, Communications & Intelligence
C4	Command, Control, Communications & Computers
CA	Conformal-array Aerial
CA	Controller Aircraft
CA	Corrective Action
CA	Gun Cruiser (USA)
CAA	Civil Aviation Authority (UK)
CAAC	Civil Aviation Administration of China
CAAIS	Computer Aided Action Information System
CAB	Common Avionics Baseline
CAC	Chengdu Aircraft Industry Corp. (China)
CAC	Coastal Artillery Computer
CAC	Combat Air Command
CAC	Combat Assessment Capability
CAC	Computer-Aided Classification
CAC	Control & Analysis Center (USA)
CAC	Custom Armoring Corp. (USA)
CAC2	Combined Arms Command & Control programme
CACCC	Combat Air Command & Control Centre
CACM	Central American Common Market
CACS	Computer Assisted Command System
CAD	Chemical Agent Decontaminant
CAD	Close-in Air Defence
CAD	Collective Address
CAD	Combat Assault Dory
CAD	Computer-Aided Design
CAD	Computer-Aided Detection
CADB	Composite Air Defense Battalion
CADE	Combined Allied Defence Experiment
CADE	Computer-Aided Design Evaluation
CADES	Computer-Aided Design Evaluation System
CADF	Commutated Aerial Direction Finder
CADM	Clustered Airfield Defeat / Dispersed Munition
CADMAT	Computer-Aided Design, Manufacture And Test
CADS	Chemical Agent Decontamination Simulant
CADS	Chemical Agent Detection System
CADS	Concept And Design Study
CADS	Controlled Aerial Delivery System
CADS	Cushion-Augmentation Device

CADWS	Close-Air Defence Weapon System
CAD/CAM	Computer-Aided Design & Manufacture
CAE	Common Applications Environment
CAE	Computer-Aided Engineering
CAEPE	Centre d'Achevement et d'Essais des Propulseurs et Engins (France)
CAF	Canadian Armed Forces. (Canadian Forces)
CAF	Changhe Aircraft Factory (China)
CAF	Citizens Air Force (South Africa)
CAF	Confederate Air Force (USA)
CAFATC	Canadian Air Transport Command (Canadian Air Force)
CAFDA/DA	Commandement Air des Forces de Defense Aerienne / Commandement de la Defense Aerienne (France)
CAFGU	Citizens Armed Force Geographical Unit. Paramilitary reserves (Philippines)
CAG	Canadian Air Group
CAG	Carrier Air Group
CAI	Close Approach Indicator
CAI	Computer-Aided Instruction
CAIA	Confederation Amicale des Ingenieurs de l'Armement (France)
CAINS	Carrier Aircraft Inertial Navigation System
CAL	Canadian Arsenals Limited
CAL	Computer-Assisted Learning
CALFFAA	Centro de Apoyo Logistico de la Fuerza Armada (Honduras)
CALHAVEN	Callington Haven Pty Ltd (Australia)
CALIB	Compact Air-Launched Ice Beacon
CALM	Crane Attachment Lorry-Mounted
CALS	Computer-Aided Logistics Support system
CAM	Chemical Agent Monitor (UK)
CAM	Chemical Agent Munition
CAM	Cockpit-Angle Measure (crew field of view)
CAM	Commerciale Avio Marina Srl (Italy)
CAM	Computer-Aided Manufacturing
CAM	Conventional Attack Missile
CAMAD	Computer-Aided Magnetic Anomaly Detection
CAMBS	Command Activated Multi-Beam Sonobuoy
CAMDS	Chemical Agent Munitions Disposal System
CAMEL	Cartridge Active Miniature Electro-Magnetic
CAMIVA	Constructeurs Associes de Materiels d'Incendie, Voirie, Aviation (France)
CAML	Cargo Aircraft Minelayer
CAMPS	Computer Assisted Message Processing System (NATO)
CAMS	Combat Aviation Management System
CAMU	Central Audio Management Unit
CANE	Command And Navigation Equipment (naval tactical data handling)
CANTASS	Canadian Towed Array Sonar System
CAO	Contract Administration Office
CAP	Carrier Air Patrol
CAP	Centre Aeroporte de Toulouse (France)
CAP	Chloroacetophenone. Tear gas
CAP	Combat Air Patrol

CAP	Combat Ammunition Production (USAF)
CAP	Combustible Augmented Plasma
CAP	Control & Audio Panel
CAP	Conventional Armaments Plan (NATO)
CAPES	Capability Evaluation System
CAPS	Conventional Armaments Planning System (NATO)
Capt	Captain
CAR	Civil Airworthiness Regulation
CAR	Conformal-Array Radar
CARA	Combined Altitude Radar Altimeter
CARABAS	Coherent All Radio Band Sensing (Sweden)
CARDS	Computer-Assisted Radar Display System
CARDS	Conceptual Architecture Design Studies
CARE	Co-operative for American Relief to Everywhere
CAREME	Centre Automatique de Reception et d'Emission de Messages (France)
CARES	Cratering And Related Effects Simulation
CARICOM	Caribbean Community
CARIFTA	Caribbean Free Trade Association
CARM	Chemical Agent Resistant Material
CARMS	Chemical Agent Remote Monitor System
CARRFA	Computer-Aided Radio Relay Frequency Assignment system
CART	Combat Aircraft Repair Team
CAS	Callibrated Airspeed
CAS	Casualty Assessment System
CAS	Chief of the Air Staff
CAS	Close Air Support
CAS	Collision Avoidance System
CAS	Combat Advice System
CAS	Combined Antenna System
CAS	Contract Administration Service
CAS	Crisis Action System (USA)
CASA	Construcciones Aeronauticas SA (Spain)
CASAP	Canadian Submarine Acquisition Programme
CASCAD	Close Air Support Cargo Dispenser
CASCOM	Combined Arms Support Command (USA)
CASD	Centro Alti Studi per la Difesa (Italy)
CASEVAC	Casualty Evacuation (UK)
CASEX	Combined Anti-Submarine Exercise
CASF	Composite Air Strike Force (USA)
CASMU	Consorzio Armamenti Spendibili Multi Uso (Italy)
CASOM	Conventionally Armed Stand-Off Missile (UK)
CASREP	Casualty Report (UK)
CASS	Camouflage Support System
CASS	Command Active Sonobuoy System
CASS	Consolidated Automatic Support System
CASS	Close Air Support System
CAST	Canadian Air-Sea Transportable Brigade
CAST	Combined Arms Staff Trainer
CAST	Computer Assisted Supervision of Transmission

CAST	Computer-Aided Search Technique
CASTOR	Corps Airborne Stand-Off Radar
CASU	Compact Air Supply Unit
CASWS	Close Air Support Weapon System
CAT	Chemical Agents Tracer
CAT	Combat Aircraft Technology
CATAC	Command Aerienne Tactique. Fighter Command (France)
CATCC	Carrier Air Traffic Control Centre
CAT-D	Chemical Agent Training Dispenser
CATE	Computer-Assisted Threat Evaluation
CATFAE	Catapult-launched Fuel-Air Explosive
CATIC	China National Aero-Technology Import & Export Corp.
CATIES	Combined Arms Training Integrated Evaluation System
CATM	Chemical Agent Training Mixture
CATRIN	Sistema Campale di Trasmissione ed Informazioni (Italy)
CATS	Comator Advanced Training System (Sweden)
CATT-B	Component Advanced Technology Test-Bed (USA)
CATTS	Combined Arms Tactical Training Simulator
CAT/FCS	Command Adjusted Trajectory / Fire Control System (USA)
CAT/LCV	Combined Arms Team / Lightweight Combat Vehicle (USA)
CAU	Central Alarm Unit
CAWC	Combined Air-Warfare Course
CAWS	Cannon Artillery Weapon Systems (USA)
CAWS	Close Assault Weapon System
CAWSEM	Countermeasures And Weapon Systems Engagement Model programme (UK)
CB	Battle cruiser
CB	Chemical-Biological
CB	Confined to Barracks
CBC	Campanhia Brasileira de Cartuchos
CBD	Commerce Business Daily (USA)
CBDE	Chemical & Biological Defence Establishment (UK)
CBF	Commander British Forces
CBG	Carrier Battle Group (US Navy)
CBI	Confederation of British Industry
C-BITE	Continuous Built-In Test Equipment
CBLS	Carrier, Bomb, Light Store
CBM	Confidence Building Measure
CBPS	Chemically & Biologically Protective Shelter
CBR	California Bearing Ratio
CBR	Chemical, Biological, Radiological warfare
CBR	Common Bomb Rack
CBRN	Caribbean Basin Radar Network
CBRN-E	Caribbean Basin Radar Network, Extended programme
CBS	Corps Battle Simulation
CBT	Computer-Based Trainer
CBTE	Carrier / Conventional, Bomb Triple Ejector
CBU	Cluster Bomb Unit
CBW	Chemical & Biological Warfare
C-C	Carbon-Carbon

CC	Charge Crueze. Shaped charge
CC	Close Combat
CC	Closed Circuit
CC	Coastal Command (Royal Air Force)
CC	Composite Command (USAF)
CC	Corps Consulaire (France)
CC3	Counter-C3
CCA	Captain's Combat Aid
CCA	Carrier-Controlled Approach
CCA	Contamination Control Area
CCAI	Canadian Centre for Advanced Instrumentation
CCAT	Carrier Control Approach Trainer
CCATTD	Common Chassis Advanced Technology Transition Demonstrator
CCB	Communications & Control Board, SHAPE (NATO)
CCC	Canadian Commercial Corporation
CCC	Combat Control Centre
CCC	Combustible Cartridge Case
CCC	Combustible Case Charge
CCC	Combustible Charge Container
CCC	Crisis Co-ordination Center (US Secretary of Defense)
CCCC	Cross Channel Co-ordination Centre
CCCES	Combat Control Concept Evaluation System
CCD	Camouflage, Concealment & Deception
CCD	Charge Coupled Device
CCD	Closed-Cycle Diesel engine technology
CCDU	Control, Communication & Display Unit
CCE	Commercial Construction Equipment
CCE	Communications Control Equipment
CCEML	Cannon-Caliber Electro-Magnetic Launcher (USA)
CCF	Combined Cadet Force (UK)
CCH	Close Combat Helicopter
CCI	Communication Interface to fire direction system (Israel)
CCIL	Continuously Computed Impact Line
CCIP	Continuously Computed Impact Point
CCIR	Comite Consultatif International des Radiocommunications (France)
CCIRM	Collection, Co-ordination & Intelligence Requirements Management
CCIS	Command & Control Information System
CCITT	International Telegraph & Telephone Consultative Committee
C-CLAW	Close Combat Laser Weapon
CCM	Conventional Cruise Missile
CCP	Central Control Point
CCP	Coherent Countermeasures Processor
CCP	Combat Command Post
CCP	Computer Control Panel
CCP	Contingency Communications Package
CCP	Control & Correlation Processor
CCPC	Civil Communications Planning Committee
CCPDS	Command & Control Processing & Display System
CCR	Computer-Controlled Range
CCRP	Continuously Computed Release Point

CCS	Central Casualty Section
CCS	Combat Control System
CCS	Command & Control Station
CCS	Communications Control System
CCS	Conformal Countermeasures System
CCS	Control & Communications Subsystem
CCS	Course Corrected Shell
CCSA	Command & Control Support Agency (DAMO) (USA)
CCSSA	Combined Cryptographical Shore Support Activity
CCT	Combat Control Team
CCT	Combat Crew Trainer
CCTA	Central Computer & Telecommunications Agency
CCTAP	Crew Compartment Test & Alarm Panel Assembly for fire fighting systems
CCTC	Command & Control Technical Center (USA)
CCTT	Close Combat Tactical Trainer (US Army)
CCTV	Closed-Circuit Television System
CCTW	Combat Crew Training Wing (USAF)
CCTWT	Coupled Cavity Travelling Wave Tube
CCU	Central Control Unit
CCU	Certificate of Clearance for Use
CCU	Cockpit Control Unit
CCU	Combat Control Unit
CCU	Common Control Unit
CCU	Communication Control Unit
CCV	Close Combat Vehicle
CCV	Control Configured Vehicle
CCV-L	Close Combat Vehicle, Light
CCW	Controller & Communications Workstation
CD	Civil Defence
CD	Compact Disc
CD	Concept Development
CD	Conference on Disarmament
CD	Corps Diplomatique (France)
CDA	Compound Document Architecture
CDAA	Circularly Disposed Antenna Array
CDB	Cast Double Base
CDBP	Command Data Buffer Program (USAF)
CDC	Command & Control Centre (France)
CDCN	Controllerate of Defence Communications Network
CDE	Chemical Defence Establishment
CDE	Conference on Disarmament in Europe
CDEC	Chief of Defence Equipment Collaboration
CDF	Ciskei Defence Force
CDFR	Commercial Demonstration Fast Reactor
CD-I	Compact Disc – Interactive
CDI	Chief of Defence Intelligence
CDI	Collector Diffusion Isolation
CDI	Conventional Defence Improvements
CDI	Course Deviation Indicator

CDIP	Continuously Displayed Impact Point
CDIRRS	Control & Display of Infra-Red Reconnaissance System
CDL	Control Data Ltd (UK)
CDM	Conditioned Diphase Modulation
CDMA	Code Division Multiple Access (data links)
CDN	Canadian
CDP	Canadian Dry Pin
CDP	Chief of Defence Procurement
CDP	Countermeasures Dispenser Pod
CDP	Critical Decision Point
Cdr	Commander
CDR	Critical Design Review
CDRB	Canadian Defense Research Board
Cdre	Commodore
CDRL	Contract Data Requirements List
CD-ROM	Compact Disc – Read Only Memory
CDRS	Container Design Retrieval System (USAF)
CDS	Chief of Defence Staff
CDS	Civil Direction of Shipping
CDS	Combat Direction System
CDS	Continuing Design Services
Cdt	Cadet
CDU	Cockpit Display Unit
CDU	Commander Display Unit
CDU	Control & Display Unit
CD-V	Compact Disc – Video
CDV	Civil Defence Volunteers
CE	Chemical Energy
CE	Concurrent Engineering
CE	Corps of Engineers Command (US Army)
CEAC	Committee for European Airspace Co-ordination
CEAM	Centre d'Experiences Aeriennes Militaires (France)
CEAT	Centre d'Essais Aeronautiques de Toulouse (France)
CEB	Combined Effects Bomblets
CEC	Commission of the European Committees
CEC	Co-operative Engagement Capability
CECAF	Centro Cartografico y Fotografico (Spain)
CECOM	US Army Communications & Electronics Command
CECORE	Centre de Commandements du Reseau (France)
CEDAM	Combined Electronic Display & Map
CEDOCAR	Centre de Documentation de l'Armement (France)
CEE	Combat Explacement Excavator
CEESIM	Combat Electromagnetic Environment Simulator
CEF	Canadian Expeditionary Force
CEFA	Chaudronnerie et Forges d'Alsace (France)
CEFACA	Centro de Catalogacao das Forcas Armadas (Portugal)
CEFO	Combat Equipment Fighting Order (UK)
CEI	Chartered Electronics Industries Pte Ltd (Singapore)
CEIEC	China National Electronics Import & Export Corp.
CEIS	Compagnie pour l'Electronique, l'Informatique et les Systemes (France)

CEL	Centre d'Essais des Landes (France)
CEL	Contractor Experience List
CELAR	Centre d'Electronique de l'Armement (France)
CELT	Combined Emitter Locator Testbed
CELT	Crypto Equipment for Low speed Telegraphy
CELTICS	Commander Extended Link for Thermal Imaging Combat Sight
CEM	Centre d'Essais de la Mediterranee (France)
CEM	Combined Effects Munition
CEM	Continous Emissions Monitor
CEMO	Combat Equipment Marching Order (UK)
CEN	C.E. Niehoff & Co. (USA)
CEN	European Committee for Standardisation
CENELEC	European Committee for Electro-Technical Standardisation
CENTAF	Central Air Force (NATO)
CENTAG	Central Army Group (NATO)
CENTCOM	Central Command (USA)
CENTECH	Century Technologies, Inc. (USA)
CENTLANT	Central Sub-Area Eastern Atlantic Command (UK)
CENTO	Central Treaty Organisation
CEO	Comprehensive Electronic Office software package
CEOA	Central Europe Operating Agency
CEOD	Chemical Explosive Ordnance Disposal
CEOI	Communications & Electronics Operating Instructions (USA)
CEOT	Communication Equipment Operator Trainer
CEP	Circular Error of Probability
CEP	Concept Evaluation Program
CEP	Contact Evaluation Plot
CEP	Contextual Enhancement Processor
CEPA	Common European Priority Areas
CEPAS	Common European Priority Areas
CEPT	Cockpit Emergency & Procedures Trainer
CER	Communication Equipment Room
CERES	Computer Enhanced Radio Emission Surveillance (UK)
CERN	Centre Europeen de Recherches Nucleaires. European Nuclear Research Centre
CERN	Centre / Conseil Europeenne pour la Recherche Nucleaire
CERT	Centre d'Etudes et de Recherches de Toulouse (France)
CES	Combat Effects Simulator
CES	Complete Equipment Schedule (UK)
CESE	Communications Equipment Support Element
CESEDEN	Centro Superior de Estudios de la Defensa Nacional (Spain)
CESG	Communications Electronics Security Group
CET	Central European Time
CET	Combat Engineer Tractor
CETIS	Centre de Transformations des Informations Scientifiques (France)
CEU	Computer Electronics Unit
CEV	Centre d'Essais en Vol. Flight test centre (France)
CEV	Combat Engineer Vehicle
CEW	Continental Early Warning

CEWI	Combat Electronic Warfare Intelligence
CEWS	Communications Electronic Warfare System
CEZUS	Compagnie Europeenne du Zirconium (France)
CF	Chaplain to the Forces
CF	Controlled Fragmentation
CFA	Communaute Financiere d'Afrique. African Financial Community
CFAC	Constant-Frequency Alternating Current
CFAR	Constant False Alarm Rate
C-FAST	Counter-Force Autonomous Surveillance & Targetting programme (USAF)
CFB	Canadian Forces Base
CFC	Central Fire Control
CFCC	Canadian Forces Communication Command
CFD	Chaff / Flare Dispenser
CFDIU	Centralised Fault Display Interface Unit
CFDS	Central Fault Display System
CFE	Canadian Forces Europe
CFE	Central Fighter Establishment (Royal Air Force)
CFE	Conventional Forces in Europe
CFEI	Cie Francaise d'Electrothermie Industrielle (France)
CFF	Chemische Fabrik Freising GmbH (Germany)
CFI	Chartered Firearms Industries Pte Ltd (Singapore)
CFI	Chief Flying Instructor
CFIT	Controlled Flight Into Terrain
CFLO	Canadian Liaison Officer
CFM	Crane, Field, Medium
CFMO	Command Flight Medical Officer
CFR	Code of Federal Regulations (USA)
CFRP	Carbon Fibre Reinforced Plastic
CFS	Central Flying School (Royal Air Force)
CFS	Chief of Fleet Support
CFSME	Canadian Forces School of Military Engineering
CFSO	Command Flight Safety Officer
CFV	Cavalry Fighting Vehicle (USA)
CG	Centre of Gravity
CG	Guided Missile Cruiser (USA)
CG	Phosgene. Chemical warfare choking agent
CGA	Compagnie Generale d'Automatisme (France)
CGCC	Centre of Gravity Control Computer
CGCSS	Commander & Gunner Crew Station Simulator
CGEN	Character Generator
CGF	Central Group of Forces. Former Soviet Forces in former Czechoslovakia (NATO)
CGG	Compagnie Generale de Geophysique (France)
CGH	Guided missile / helicopter cruiser
CGHA	Compressed Gas Association (USA)
CGI	Computer-Generated Image
CGM	Computer Graphics Metafile
CGN	Guided Missile Cruiser, nuclear propulsion (USA)
CGRM	Commandant General, Royal Marines

CGS	Central Gunnery School (Royal Air Force)
CGS	Chief of the General Staff
CGS	Crew Gunnery Simulator
CGV	Computer-Generated Voice
CGWIC	China Great Wall Industry Corp.
CHA	Cylindrical Hydrophone Array
CHAALS	Communications High Accuracy Airborne Location System
CHAIR	Control Handling Aid for Increase Range
CHALS-X	Communication High-Accuracy Location System – Exploitable (USA)
CHAMP	Configurable Hardware Algorithm Mappable Preprocessor
CHANCOM	Channel Command (NATO)
CHAPS	Chamberlain Armor Protection System (USA)
CHARM	Challenger Chieftain Armament
CHARM	Composite High Altitude Radiation Model
CHASE	Chemical Agents Sensor
CHENG	Office of the Chief Engineer of the Navy (USA)
CHIP	Challenger Improvement Programme
CHM	Civilian Hood Mask
CHOD	Chief Of Defence (NATO)
CHODS	Chiefs Of Defence (NATO)
CHOP	Change of OPCON
CHOTS	Corporate Headquarters Office Technology System (UK)
CHP	Combined Heat & Power
CHS	Common Hardware System
CI	Channel Islands
CI	Chief Instructor
CI	Compression Ignition
CI	Configuration Item
CI	Control Indicator
CI	Counter-Intelligence
CIA	Central Intelligence Agency (USA)
CIC	Combat Information Centre
CICAT	Center for Industry Cooperation & Trade (USA)
CIC-PT	Combat Information Centre Procedure Trainer
CID	Charge-Injection Device
CID	Commander's Integrated Display
CIDA	Co-ordinating Installation Design Authority
CIDIS	Combat Information Display System
CIDS	Collateral Information Display Station
CIEG	Common Information Exchange Glossary
CIEL	Common Information Exchange Language
CIG	Commerce International Group Ltd (Spain)
CIG	Communications Industries Group (Iran)
CIGARS	Console Internally Generated And Refreshed Symbols programme (USA)
CIGS	Chief of the Imperial General Staff. Now CGS
CIGTF	Central Inertial Guidance Test Facility (USAF)
CII	Communications Instruments Inc. (USA)
CII	Communications International, Inc. (USA)
CIJ	Close-In Jamming

CIJ	Cour International de Justice
CILAS	Compagnie Industrielle des Lasers (France)
CIM	Computer-Integrated Manufacture
CIMD	Centre D'Identification des Materiels de la Defense (France)
CIMIC	Civil Military Co-operation
CINA	Commission Internationale de la Navigation Aerienne
CinC	Commander-in-Chief
CINC	Commander-In-Chief
CinC UKAIR	Commander-in-Chief, UK Air Forces
CINCEASTLANT	Commander-in-Chief, Eastern Atlantic (UK)
CINCENT	Commander-in-Chief, Allied Forces Central Europe (NATO)
CINCEUR	Commander-in-Chief, European Command
CINCFLEET	Commander-in-Chief, Fleet (Royal Navy)
CINCGERFLT	Commander-in-Chief, German Fleet (NATO)
CINCHAN	Commander-in-Chief, Channel (NATO)
CINCHEL	Commander-in-Chief, Channel & Eastern Atlantic Areas (NATO)
CINCIBERLANT	Commander-in-Chief, Iberian Atlantic Area (NATO)
CINCLANT	Commander-in-Chief, Atlantic Command
CINCLANTFLT	Commander-in-Chief, Atlantic Fleet (NATO)
CINCNAVHOME	Commander-in-Chief, Naval Home Command (Royal Navy)
CINCNORTH	Commander-in-Chief, Allied Forces Northern Europe (NATO)
CINCPAC	Commander-in-Chief, Pacific Command
CINCSAC	Commander-in-Chief, Strategic Air Command
CINCSOUTH	Commander-in-Chief, Allied Forces Southern Europe (NATO)
CINCUKAIR	Commander-in-Chief, UK Air Forces
CINCUSNAVEUR	Commander-in-Chief, US Navy, Europe (NATO)
CINCWESTLANT	Commander-in-Chief, Western Atlantic (NATO)
CINTI	The Cincinnati Gear Company (USA)
CIO	Central Imagery Office (USA)
CIOR	Interallied Confederation of Reserve Officers
CIR	Continuous Infra-Red
CIRC	Central Information Reference & Control (FASTC) (USA)
CIRCE	Cossor Interrogation & Reply Cryptographic Equipment
CIRPLS	Computer Integration of Requirements Procurement Logistics & Support
CIRRA	Compagnie Industrielle des Reflecteurs Radar (France)
CIRTEVS	Compact Infra-Red Television System
CIS	Chartered Industries of Singapore
CIS	Combat Identification System
CIS	Combined Influence Sweep
CIS	Command Information System
CIS	Command Intelligence System
CIS	Commonwealth of Independent States (former Soviet Union)
CIS	Communication & Information System
CISCEA	Commissao de Implantacao de Sistema de Controle do Espaco Aereo (Brazil)
CISD	Communications & Information Systems Division, SHAPE (NATO)
CIT	Combined Interrogator-Transponder
CITG	Commander, Initial Training Group (UK)
CITS	Centrally Integrated Test System

CITV	Commander's Independent Thermal Viewer (USA)
CIU	Control Interface Unit
CIWS	Close In Weapon System
CJCS	Chairman of the Joint Chiefs of Staff
CJMOC	Combined Joint Maritime Operations Centre (NATO)
CJS	Communication Channel & Jammer Simulator
CK	Cyanogen Chloride. Chemical warfare blood agent
CKD	Component Knocked Down (for assembly elsewhere)
CL	Chemical Laser
CL	Light cruiser
CLA	Low-level Approach Centre (France)
CLAES	Cryogenic Limb Array Etalon Spectrometer
CLAFA	Logistics & Administrative Command (Portugal)
CLAMS	Clear Lane Marking System (USA)
CLARA	Compact Laser Radar
CLASS	Closed-Loop Artillery Simulation System
CLASS	Coherent Laser radar Airborne Shear Sensor
CLASSIC	Covert Local Area Sensor System for Intruder Classification
CLAW	Close Light Assault Weapon
CLAWS	Close Combat Light Armor Weapon System (USA)
CLC	Charge Linear Cutting
CLC	Comlinear Corp. (USA)
CLC	Command Launch Computer
CLD	Camion Lourd de Depannage (France)
CLD	Communications Logistics Depot
CLDP	Convertible Laser Designation Pod
CLDS	Cockpit Laser Designation System
CLES	Comite de Liaison des Exposants de Satory (France)
CLEWP	Cleared Lane Explosive Widening Path Charge
CLGE	Cannon-Launched Guidance Electronics
CLGP	Cannon-Launched Guided Projectile (USA)
CLNS	Connectionless Network Service
CLO	C.L. Organization (Israel)
CLO	Logistics & Training Command (RNLAF)
CLOS	Command to Line-Of-Sight
CLRSTAR	Cueing L-band Rugged Search Track & Acquisition Radar
CLS	Capsule Launch System
CLS	Chief of Logistics Support (UK)
CLS	Contractor Logistics Support
CLT	Container Load Trailer
CLT	Light cruiser, training
CLU	Command Launch Unit
CM	Cluster Munition
CM	Cruise Missile
CMA INTL	Commercial Marketing Assistance International (USA)
CMAG	Cruise Missile Advanced Guidance
CMAS	Crisis Management ADP System
CMBG	Canadian Mechanised Brigade
CMC	Canadian Marconi Co.
CMC	Central Military Commission (China)

CMC	Cruise Missile Carrier
CMD	Countermeasures Duties
CMD	Cratering Munition Dispenser
CMDS	Countermeasures Dispenser System
CMEA	Council for Mutual Economic Assistance
CMF	Conceptual Military Framework (NATO)
CMHR	Combustion Modified Highly Resilient
CMI	Cockerill Mechanical Industries (Belgium)
CMI	Compilation, Mission support, Integration & training facility (Australia)
CMI	Computer-Managed Instruction
CMI	Cruise Missile Interface
CMIK	Cruise Missile Integration Kit
CMLSA	Commercial Microwave Landing System Avionics
CMMCA	Cruise Missile Mission Control Aircraft
Cmnd	Command Paper
CMOD	Compact Meteorological & Oceanographic Drifter
CMOS	Complementary Metal Oxide Silicon (semiconductor)
CMP	Countermeasures Precursor
CMP	Counter-Military Potential
CMP	Office of the Director, Cruise Missile Project (USA)
CMPS	Crooks Michell Peacock Stewart Pty Ltd (Australia)
CMRA	Cruise Missile Radar Altimeter
CMRS	Countermeasures Receiver System
CMRS	Crash / Maintenance Recorder System
CMRST	Committee on Manpower Resources for Science & Technology
CMS	Cockpit Management System
CMS	Combat Management System
CMS	Combat Mission Simulator
CMS	Combat Monitoring System
CMS	Control Monitor Set
CMS	Crown Metro Salchi SpA (Italy)
CMSAF	Chief Master Sargeant of the Air Force (USAF)
CMT	Cadmium Mercury Telluride (chemical abbreviation CdHgTe)
CMT	Critical Military Target
CMU	Course Measuring Unit
CMV	Combat Mobility Vehicle (USA)
CN	Chloroacetophone. Chemical warfare tear agent
CN2H	Conduit Nuit, second generation, Helicopters
CNA	Center for Naval Analyses (USA)
CNAD	Conference of National Armament Directors
CNB	Chemical warfare tear agent
CNC	Chemical warfare tear agent
CNCE	Communications Nodal Control Element
CND	Campaign for Nuclear Disarmament
CND	Captain, Naval Drafting (UK)
CNEI	C&N Electrical Industries Ltd (UK)
CNES	Centre National d'Etudes Spatiales (France)
CNI	Communications, Navigation & Identification (USA)
CNIM	Constructions Navales et Industrielles de la Mediterranee (France)

CNIU	Communications / Navigation Interface Unit
CNLP	Connectionless Network Layer Protocol
CNO	Chief Nursing Officer
CNO	Chief of Naval Operations (USA)
CNOCS	Commander, Naval Operations Control systems (NATO)
CNP	Chief of Naval Personnel & Second Sea Lord
CNR	Combat Net Radio
CNRE	Combat Net Radio Environment
CNRI	Combat Net Radio Interface
CNRS	Centre National de la Recherche Scientifique (France)
CNS	Chemical warfare tear agent
CNS	Chief of Naval Staff & First Sea Lord
CNS	Communications Network Simulator
CNS	Communications, Navigation, Surveillance
CNSA	Commodore Naval Ship Acceptance
CNSO	Chief Naval Signal Officer (UK)
CNT	Canister
CNTC	Office of the Commander, Naval Telecommunications Command (USA)
CO	Central Office
CO	Commanding Officer
CO2	Carbon dioxide
COAR	Consola del Arma. Weapon control console (Spain)
COBOL	Common Business-Oriented Language
COBRA	Counter Battery Radar
COBUMA	Defensie Codificatie Centrum Materieel (Netherlands)
COC	Combat Operation Centre
COC	Command Operations Centre
COCC	Contractor's Operational Control Center (USA)
COCOM	Co-ordinating Committee for East-West Trade Policy
COCT	Cab-Over Cargo Truck
COD	Carrier Onboard Delivery
CODAG	Combined Diesel And Gas Turbine
CODERM	Committee On Defence Equipment Reliability & Maintainability
CODLAG	Combined Diesel, Electric And Gas
CODOG	Combined Diesel Or Gas Turbine
CODSIA	Council Of Defense & Space Industry (USA)
COE	Cab Over Engine
COEC	Council Operations & Executive Committee, NATO HQ
COFAM	Computer Facility for Mine Warfare
COFDEN	Commander, Operational Forces, Denmark
CofN	Controller of the Navy
COFS	Chief Of Staff, SHAPE (NATO)
COFT	Conduct Of Fire Trainer
COGAG	Combined Gas Turbine And Gas Turbine
COGEMA	Compagnie Generale de Matieres Nucleaires (France)
COGOG	Combined Gas Turbine Or Gas Turbine
COH	Clyde Operational Headquarters
COIL	Chemical Oxygen Iodine Laser
COIN	Counter-Insurgency

COINS	Computer Operated Instrument System
COJACS	Combat Jamming / Communication Support System
COJAS	Coherent Jammer Simulator
Col	Colonel
COLDS	Common Opto-electronic Laser Detection System
COLOC	Change Of Locations Of Command (NATO)
COLT	CO2 Laser Technology
Com	Commander
Com	Commodore
COM	Compact Optronic Mast
COMAAFCE	Commander, Allied Air Forces, Central Europe (NATO)
COMAIRBALTAP	Commander, Allied Air Forces, Baltic Approaches (NATO)
COMAIRCHAN	Allied Maritime Air Forces Command Channel (UK)
COMAIRNON	Commander Air, Northern Norway (NATO)
COMAIRSONOR	Commander, Allied Air Forces, South Norway (NATO)
COMAIRSOUTH	Commander, Allied Air Forces, Southern Europe (NATO)
COMAMF (L)	Commander, ACE Mobile Forces, Land (NATO)
COMAR	Comando General de la Armada (Uruguay)
COMARAIRMED	Commander, Maritime Air Forces, Mediterranean (NATO)
COMASWSTRIKFOR	Commander, ASW Strike Force (NATO)
COMBALTAP	Commander, Allied Forces, Baltic Approaches (NATO)
COMBAT-SIM	Computerized Battle Simulation
COMBENECHAN	Commander, Benelux Sub-Area, Channel (NATO)
COMCEN	Communications Centre
COMCENTAG	Commander, Central Army Group (NATO)
COMCENTLANT	Commander, Central Atlantic Sub-Area (NATO)
COMCLYDE	Commander, Clyde Area (NATO)
COMDEF	Common Defense Exhibition & Seminar
Comdr	Commander
Comdt	Commandant
COMECON	Council for Mutual Economic Assistance
COMED	Combined Map & Electronic Display
COMEDCENT	Commander, Central Mediterranean (NATO)
COMEDEAST	Commander, Eastern Mediterranean (NATO)
COMEDNOREAST	Commander, Northeast Mediterranean (NATO)
COMFAIRMED	Commander, Fleet Air Mediterranean (Italy)
COMFIVEATAF	Commander, 5th Allied Tactical Air Force (NATO)
COMFOURATAF	Commander, 4th Allied Tactical Air Force (NATO)
COMFUSNA	Commando de Fusileros Navales (Uruguay)
COMGEIM	Commandancia General de la Infanteria de Marina (Spain)
COMGIBMED	Commander, Gibraltar Mediterranean (NATO)
COMGRNORSEA	Commander, German North Sea (NATO)
COMHELFLEET	Hellenic Fleet Command (Greece)
COMICEDEFOR	Island Commander Iceland & Commander Fleet Air Keflavik (Iceland)
COMINT	Communications Intelligence
COMJAM	Common Jammer
COMLANDJUT	Commander, Allied Land Forces, Jutland (NATO)
COMLANDNON	Commander, Allied Land Forces, North Norway (NATO)
COMLANDSONOR	Commander, Allied Land Forces, South Norway (NATO)

COMLANDSOUTH	Commander, Allied Land Forces, Southern Europe (NATO)
COMLANDSOUTHEAST	Commander, Allied Land Forces, Southeastern Europe (NATO)
COMLAND-ZEALAND	Commander, Allied land Forces, Zealand (NATO)
COMM	Communications
COMMAIR (CHEL)	Commander Maritime Air, (Channel, East Atlantic) (NATO)
COMMAIRCENTLANT	Commander Maritime Air, Central Atlantic (NATO)
COMMAIRCHAN	Commander Maritime Air, Channel (NATO)
COMMAIREASTLANT	Commander Maritime Air, Eastern Atlantic (NATO)
COMMAIRNORLANT	Commander Maritime Air, Northern Atlantic (NATO)
COMMAIRPLYMCHAN	Commander Maritime Air, Plymouth Channel (NATO)
COMMCEN	Communications Centre
COMMW	Commander, Mine Warfare / War vessels (NATO)
Commy	Commissary
COMNAEWF	Commander, NATO Airborne Early Warning Force (NATO)
COMNAVACTUK	Commander, US Naval Activities, United Kingdom
COMNAVBALTAP	Commander, Allied Naval Forces, Baltic Approaches (NATO)
COMNAVNON	Commander, Allied Naval Forces, North Norway (NATO)
COMNAVSONOR	Commander, Allied naval Forces, South Norway (NATO)
COMNAVSOUTH	Commander, Allied Naval Forces, Southern Europe (NATO)
COMNON	Commander, Northern Norway (NATO)
COMNORECHAN	Commander, Nore Sub-Area, Channel (NATO)
COMNORECHAN	Nore Sub-Area Command (UK)
COMNORLANT	Commander, North Atlantic Sub-Area (NATO)
COMNORTHAG	Commander, Northern Army Group (NATO)
Comp B	Composition B
COMPLYMCHAN	Commander, Plymouth Sub-Area, Channel (NATO)
COMPUSEC	Computer Security
COMPUSEC RA	Computer Security Risk Analysis
COMSEC	Communications Security
COMSERVGRU	Commander, Service Group (NATO)
COMSIM	Communications Simulator
COMSIXATAF	Commander, 6th Allied Tactical Air Force (NATO)
COMSONOR	Commander, Southern Norway (NATO)
COMSTANAVFORCHAN	Commander, Standing Naval Force Channel (NATO)
Comsteel	Commonwealth Steel Company Ltd (Australia)
COMSTRIKEFORSOUTH	Commander, Naval Striking & Support Forces, Southern Europe (NATO)
COMSTRIKFLTLANT	Commander, Strike Fleet Atlantic (NATO)
COMSUB (CHELL)	Commander, Submarines (Channel, East Atlantic) (NATO)
COMSUBACLANT	Commander, Submarines Allied Command, Atlantic (NATO)
COMSUBEASTLANT	Commander, Submarines Forces, Eastern Atlantic Area (NATO)
COMSUBLANT	Commander, Submarine Forces Atlantic (USA)
COMSUBMED	Commander, Submarines Mediterranean (NATO)
COMTEL	Commercial Telecommunications Corp. (USA)
COMTWOATAF	Commander, 2nd Allied Tactical Air Force (NATO)
COMVAT	Combat Vehicles Armament Technology (USA)
CONAS	Combined Nuclear And Steam
CO-NEWS	Communications Naval Electronic Warfare System

CONFORMISC	Etat Major du Commandement des Formations (France)
CONMAROPS	Concept of Maritime Operations (NATO)
CONTAC	Consola Gestion Tactica. Fire control console (Spain)
CONUS	Continental United States (excluding Hawaii)
COOP	Craft-Of-Opportunity. Mine detection technique
COPEX	Covert & Operational Procurement Exhibition
CORAD	Co-ordinated Roland Air Defence
CORAS	Future correctable ammunition (Netherlands)
Corp	Corporal
CORPS SAM	Corps Surface-to-Air Missile system (US Army)
CoS	Chief of Staff
COSAG	Combined Steam And Gas Turbine
COSHH	Control Of Substances Hazardous to Health
COSMIC	Combat Systems Multi-format Information Centre
COSPAR	Committee for Space Research
COSSEC	Chief(s) of Staff Secretariat
COSTIND	Commission of Science Technology & Industry (China)
COSYS	Combat System. Naval command & weapon control system
COTAC	Conduite de Tir Automatique pour Char. Tank automatic fire control system (France)
COTAM	Commandement du Transport Aerien Militaire (France)
COTS	Commercial Off-The-Shelf software
COTSAA	Centre d'Operations des Transports (France)
COTT	Control Of Transmission Test
COV	Counter Obstacle Vehicle (USA)
Coy	Company
cp	Controllable pitch (propeller)
CP	Circulaire Pivot. Circular pivot (France)
CP	Command Post
CP	Concrete-Piercing
CP2	Contractor Performance Certification Program (USA)
CPA	Cabin Public Address
CPA	Closest Point of Approach
CPA	Continuous Patrol Aircraft
CPACS	Coded Pulse Anti-Clutter System
CPA/SG	Central Pulse Amplifier / Symbol Generator
CPC	Conflict Prevention Centre
CPCN	Command Post Communication Network
CPF	Canadian Patrol Frigate
CPG	Co-Pilot / Gunner
CPI	Communications Processor & Interface
CPJI	Cour Permanente de Justice Internationale (France)
Cpl	Corporal
CPL	Characters Per Line
CPM	Centre de Programmation de la Mer (France)
CPM	Civil, Police, Military activities (Thailand)
CPM	Control Processor Module
CPMIEC	China Precision Machinery Import & Export Corp. (China)
CPM/P	Command Post Modem / Processor
CPO	Chief Petty Officer (Royal Navy)

CPOP	Command Post / Observation Post
CPP	Controlled Pitch Propeller
CPR	Cardio-Pulmonary Resuscitation
CPR	Coherent-Pulse Radar
CPR	Covert Penetration Radar
CprM	Campo de Provas da Marambaia. Marambaia proving grounds (Brazil)
CPS	Covert Penetration System
CPT	Cockpit Procedures Trainer
CPU	Central Processing Unit
CPU	Communications Processing Unit
CPV	Command Post Vehicle
CPX	Command Post Exercise (UK)
CQB	Close Quarter Battle
CQBR	Close Quarter Battle Range
CQE	Close Quarter Engagements
CR	Central Region
CR ARRV	Challenger Armoured Repair & Recovery Vehicle (UK)
CRAF	Civil Reserve Air Fleet
CRANMOS	Cranfield Motion Systems Div. (UK)
CRARRV	Challenger Armoured Repair & Recovery Vehicle (UK)
CRB	Capsill Roller Beam
CRC	Coleman Research Corp. (USA)
CRC	Container Research Corp. (USA)
CRC	Control & Reporting Centre
CRCU	Crypto Remote Control Unit
CRDEC	Chemical Research, Development & Engineering Center
CRE	Communications Radar Exciter
CREST	Committee of Scientific & Technological Research
CREST	Comprehensive Radar Effects Simulator Trainer
CRI	Computer Resources International A/S (Denmark)
CRIAP	Central Region Initial ACCS Programme (Netherlands)
CRIS	Coastal Radar Integration System (Denmark)
CRIS	Conception, Realisation et Implantation de Systemes (France)
CRISP	Computer Reconstructed Images from Scene Photographs
CRITCOM	Critical intelligence Communications (USA)
CRL	Common Rail Launcher
CRMP	Corps of Royal Military Police
CROWCASS	Criminals of War Commission
CRP	Control & Reporting Post
CRPA	Controlled Reception Pattern Antenna
CRR	Carro de Reconhecimento Sobre Rodas. Reconnaissance tracking scout car
CRRC	Combat Rubber Raiding Craft
CRRES	Combined Release & Radiation Effects Satellite
CRT	Cathode Ray Tube
CRT	Component Repair Technologies Inc. (USA)
CRTE	Combat Rescue Training Exercise
CRU	Control Radio Unit
CRW	Counter Revolutionary Warfare

CS	Central Services division (UK)
CS	Cone-Stabilised
CS	Confined Space
CS	O-chlorobenzamalonitrile. Chemical warfare tear agent
CSA	Canadian Space Agency
CSA	Canadian Standards Association
CSA	Chief Scientific Adviser, defence scientific staff (UK)
CSA	Computer Sciences of Australia Pty Limited
CSA	Computer System Architects Ltd (UK)
CSAF	Chief of Staff (USAF)
CSAR	Combat Search And Rescue
CSAS	Command & Stability Augmentation System
CSB	Combat Support Boat
CSBM	Confidence & Security Building Measures
CSC	Cable Supply & Consulting Co. Pte Ltd (Singapore)
CSC	Communication System Control
CSC	Compass System Controller
CSC	Computer Sciences Corp. (USA)
CSC	C.S. Components Ltd (UK)
CSCE	Communications Systems Control Element
CSCE	Conference on Security & Co-operation in Europe
CSCG	Communications Security Control Group (USA)
CSCM	Conference on Security & Co-operation in the Mediterranean & Middle East
CSCS	Contractor's Satellite Control Site (USA)
CSD	Common Strategic Doppler
CSD	Constant-Speed Drive
CSD	Contemporary Systems Design Ltd (UK)
CSDT	Control for Submarine Discharge Torpedo
CSEDS	Combat Systems Engineering Development Site
CSEU	Confederation of Shipbuilding & Engineering Unions
CSF	Combined Service Forces
CSG	Contact Signature Generator
CSG	Control Site Group (radar control & communications adaption)
CSgt	Colour Sergeant
CSH	Combat Support Helicopter
CSH	Combat System Highway interface
CSI	Circuiti Stampati Italia
CSI	Computer Synthesised Image
CSICDB	Czechoslovak Independent Chemical Defence Battalion (Czechoslovakia)
CSICS	Combatant Ships Integrated Communications System
CSIT	Computer Science & Information Laboratory (USA)
CSL	Light cargo ship
CSLT	Control for Surface Launched Torpedoes
CSM	Combat Support Module
CSM	Continental Shelf Mine
CSMA/CD	Carrier Sense Multiple Access / Collision Detection network
CSMS	Corps Support Missile System
CSMU	Crash Survivable Memory Unit

CSO	Chief Staff Officer, Reserves (Royal Navy)
CSO	Command Signals Officer (Royal Air Force)
CSOC	Consolidated Space Operations Centre
CSR	Covert Strike Radar
CSRA	Combat System Requirements Analysis
CSRDF	Crew Station Research & Development Facility
CSRL	Common Strategic Rotary Launcher
CSS	Chinese Surface-to-Surface missile
CSS	Combat Service Support
CSS	Command Support System (UK)
CSS	Computer Sighting System
CSS	Computer Support System
CSS	Communication Support Segment for JSIPS
CSSCS	Combat Service Support Control System
CSSD	Command Support System Demonstrator
CSSD	Strategic Defense Command (US Army)
CSSE	Chief Strategic Systems Executive
CSS-N	Chinese Surface-to-Surface (Navy) missile
CSSS	Combat Service Support System
CSS/DPEE	Central Scientific Services DPEE (UK)
CST	Central Standard Time
CST	Combat Support Trailer
CST	Conventional Stability Talks
CSTB	Combat System Test-Bed
CSTD	Committee on Science & Technology for Development
CSTF	Cross-Scan Terrain-Following radar mode
CSTS	Combat Simulation Test System
CSU	Command Sensor Unit
CSU	Communications Switching Unit
CSU	Control Selection Unit
CSV	Close-range Surveillance system
CSW	Conventional Stand-off Weapon
CSWG	Communications Systems Working Group
CSWS	Corps Support Weapon System
CTA	Case Telescoped Ammunition
CTA	Circuit Technology (Australia)
CTA	Conduit de Tir Analogue. Analogue fire control system (France)
CTB	Common Trailer Base
CTBT	Comprehensive Test Ban Treaty
CTC	Centralised Traffic Control
CTC	Combat Training Center
CTC	Communications Training Centre (Royal Navy)
CTD	Colour Tactical Display
CTD	Conduit de Tir Digitale. Digital fire control system (France)
CTE	Contractual Technical Evaluation
CTEX	Centro Tecnologico do Exercito. Army technological centre (Brazil)
CTF	Central Test Facility
CTF	Commander, Task Force (NATO)
CTG	Commander, Task Group (NATO)
CTG	Compact Tank Gun

CTH	Conduit de Tir Hybride. Naval fire control system (France)
CTI	Central Tyre Inflation
CTI	Comprehensive Technologies International Inc. (USA)
CTL	Cincinnati Testing Laboratories Inc. (USA)
CTMS	Compact Tactical Message Switch
CTMS	Conduit de Tir Multisenseurs. Multi-sensor fire control system (France)
CTO	Chief Technical Officer
CTOL	Conventional Take-Off & Landing
CTRA	Carro de Transporte Sobre Rodas Anfibo. Amphibious tracking scout car (Brazil)
CTRPS	Cine Target Range Projection System
CTS	Central Tactical System
CTS	Communications Training System
CTT	Challenger Training Tank
CTT	Combat Tactics Trainer
CTT	Command Tactical Trainer
CTT	Command Team Training
CTT 5	Combined Tactical Trainer, Stage 5 feasibility study
CTTO	Central Tactics & Trials Organisation (UK)
CTU	Control Terminal Unit
CTVS	Cockpit TV Sensor
CTZ	Corps Tactical Zone
CU	Control Unit
CUBE	Configuration Unitee Banalisee d'Exploitation (France)
CUCV	Commercial Utility Cargo Vehicle
CUDIXS	Common User Digital Information Exchange System (USA)
CUP	Capability Upkeep Programme
CUSRPG	Canada/US Regional Planning Group
CV	Aircraft Carrier (USA)
CV	Combat Vehicle
CV	Cryptographic Variable
CVA	Attack Aircraft Carrier
CVAM	Special Air Missions Office (USA)
CVAST	Combat Vehicle Armament System Technology (USA)
CVBG	Fleet carrier Battle Group
CVC	Combat Vehicle Crewman
CVC2	Combat Vehicle Command & Control programme (USA)
CVH	Helicopter carrier
CVL	Correlation Velocity Log
CVL	Light aircraft carrier
CVN	Aircraft Carrier, nuclear propulsion (USA)
CVR	Cockpit Voice Recorder
CVRDE	Combat Vehicle Research & Development Establishment
CVROS	Compact Video Rate Optical Scanner
CVR(T)	Combat Vehicle Reconnaissance (Tracked)
CVR(W)	Combat Vehicle Reconnaissance (Wheeled)
CVS	Aircraft Carrier, anti-submarine warfare (USA)
CVSD	Continuously Variable Slope Delta (modulation)
CVT	Controlled Variable Time

CVTTS	Combat Vehicle Thermal Targetting System
CVV	Aircraft Carrier, Medium-Sized (USA)
CVW	Carrier air wing (US Navy)
CW	Chemical Warfare
CW	Continuous Wave
CWAR	Continuous Wave Acquisition Radar
CWC	Chemical Weapons Convention
CWCD	Chemical Weapons & Civil Defence
CWCS	Common Weapons Control System
CWD	Chemical Warfare Defence
CWDD	Chemical Warfare Directional Detector
CWDD	Continuous Wave Deuterium Demonstrator
CWILL	Continuous Wave Illuminating Radar (USA)
CWP	Centralised Warning Panel
CWR	Continuous Wave Radar
CWS	Common Weapon Sight (UK)
CWS	Container Weapon System
CWS	Co-ordination & Warning System (Denmark)
CWS	Cupola Weapon Station
CWW	Cruciform Wing Weapon
CX	Phosgene Oxime. Chemical warfare blister agent
CZ	Convergence Zone (sonar)
CZMCS	Combat Zone Mobile Communication System (Turkey)
CZP	Correct Zeroing Point
C&C	Command & Control
C&FD	Coating & Filter Design (UK)
C+EE	Central & Eastern Europe (NATO)
C/A	Coarse Acquisition
c/n	Construction (or constructor's) number
C/R	Counter-Rotating (propeller)
C/Sgt	Colour Sergeant

D

D	Delay (fuzes)
DA	Defence Adviser
DA	Delayed Action
DA	Diphenylchloroarsine. Chemical warfare vomiting agent
DA	Direct Action
DA	Double-Action
DAA	Digital / Analog Adapter
DAACM	Direct Airfield Attack Combined Munition
DAAT	Digital Angle of Attack Transmitter
DAB	Defence Acquisition Board (USA)
DABM	Defence Against Ballistic Missile
DAC	Digital to Analog Converter
DAC	Discretionary Access Control
DAC	Douglas Aircraft Co. (USA)
DaCC	Display and Computer Console
DACM	Data Adapter Control Mode
dACs	Digital AC Servos
DACS	Deployable Acoustic Calibration System
DACS	Directorate of Aerospace Combat Systems (Canada)
DACS	Office of the Chief of Staff (USA)
DACT	Dissimilar Air Combat Training
DACU	Defence Arms Control Unit (UK)
DAD	Deep Air Defence
DADC	Digital Air Data Computer
DADS	Digital Air Data System
DADS	Digital Audio Distribution System
DADS	Direct Application Decontamination System (Germany)
DAFCS	Digital Automatic Flight Control System
DAIS	Digital Avionics Information System
DAISY	Digital Action Information System (Netherlands)
DALO	Divisional Air Liaison Officer (UK)
DALO	Office of the Deputy Chief of Staff for Logistics (USA)
DALP	Defence Advanced Lithography Program (USA)
DAMA	Demand Assigned Multiple Access
DAMO	Office of the Deputy Chief of Staff for Operations & Plans (USA)
DAMS	Drum Auxiliary Memory Sub-Unit
DAP	Data Access Point
DAP	Decontamination Apparatus, Portable
DAP	Distributed Array Processor
DAPS	Defence & Aerospace Publishing Services SA (Switzerland)
DAR	Defence Acquisition Regulation
DARA	German Space Agency
DARCOM	Development & Readiness Command (US Army)
DARIC	Defense Automation Resources Information Center (USA)
DARKAS	Darstellungsgerat Kanonenabschuss (Germany)
DArmRD	Directorate of Armament Research & Development (UK)
DARPA	Defense Advanced Research Projects Agency (USA)

DARS	Digital Attitude-Reference System
DART	Data And Remote control Terminal
DART	Deployable Automatic Relay Terminal
DARTS	Digital Airborne Radar Threat Simulator
DARTS	Distributed Architecture Robust Tactical System (Canada)
DAS	Defensive Aids Suite
DAS	Defensive Avionics System
DASC	Direct Air Support Central
DASCU	Digital Automotive System Control Unit
DASH	Drone Anti-Submarine Helicopter
DASP	Discrete Analog Signal Processing
DASS	Defensive Aids Subsystem
DAST	Drones for Aerodynamic & Structural testing
DAT	Direction des Armements Terrestres (France)
DATAR	Detection And Tactical Alert of Radar
DATEC	Data And Telecommunications
DAU	Data Acquisition Unit
DAU	Digital Amplifier Unit
DAW	Dedicated All-Weather aircraft
DAWN SET	Defence & Early Warning Network, Southeastern Turkey
DB	Double Base
DBB	Delayed Blowback
DBFM	Defensive Basic Flight Manoeuvres
DBGF	Database Generation Facilities
DBMS	Database Management System
DBNS	Doppler Bombing / Navigation System
DBP	Draw Bar Pull
DBS	Direct Broadcasting Satellite
DBS	Doppler Beam Sharpening
DC	Depth Charge
DC	Detection Centre for missile system
DC	Digital Computer
DC	Diphenylcyanoarsine. Chemical warfare vomiting agent
DCA	Defence Codification Authority (UK)
DCA	Defence Co-operation Agreement
DCA	Defense Communications Agency (USA)
DCA	Defense Contre Avions. Anti-aircraft defence (France)
DCA	Defensive Counter-Air
DCAe	Direction des Constructions Aeronautiques (France)
DCAN	Direction des Constructions et Arms Navales (France)
DCAP	Draft Conventional Armaments Plan
DCC	Data Communication Computer
DCC	Defence Communications Centre (UK)
DCC	Digital Computer Complex
DCC	Drone Control Centre
DCC	Dynamic Controls Corp. (USA)
DCCA	Direction Centrale du Commissariat de l'Air (France)
DCCAT	Direction Centrale du Commissariat de l'Armee de Terre (France)
DCCM	Direction Centrale du Commissariat de la Marine (France)
DCCS	Deputy Chief of Staff, Support, SHAPE (NATO)

DCDS (S)	Deputy Chief of the Defence Staff (Systems)
DCE	Data Communications Equipment
DCEC	Defense Communications Engineering Center (USA)
DCEE	Defence Components & Equipment Exhibition
DCG	Direction Centrale du Genie (France)
DCI	Directorate of Cataloging & Identification (Canada)
DCI (RN)	Defence Council Instructions (Royal Navy)
DCJSS	Direct Coupled Jamming / Surveillance System
DCMA	Defence Combined Material Agency (Norway)
DCMAA	Direction Centrale du Materiel de l'Armee de l'Air (France)
DCMAT	Direction Centrale du Materiel de l'Armee de Terre (France)
DCMS	Digital Communications Management System
DCN	Defence Communications Network
DCN	Direction des Constructions Navales (France)
DCNO	Deputy Chief of Naval Operations
DCNRI	Dismounted Combat Net Radio Interface
DCOS	Deputy Chief of Staff
DCP	Digital Communications Processor
DCPG	Defense Communications Planning Group (USA)
DCPU	Display Control Power Unit
DCR	Digitally Coded Radar
DCS	Defence Communications System (USA)
DCS	Defense Construction Service (Denmark)
DCSEA	Service des Essences des Armees Direction Centrale (France)
DCSO	Deputy Chief of Staff, Operations (NATO)
DCSO	Deputy Commander for Space Operations (USAF)
DCSS	Deputy Chief of Staff, Support (NATO)
DCS/R&D	Deputy Chief of Staff for Research & Development
DCT	Depth Charge Thrower
DCT	Digital Communications Terminal
DCU	Data Collection Unit
DCU	Detection & Control Unit
DCU	Digital Computer Unit
DCU	Dispenser Control Unit
DCU	Display / Controller Unit
DD	Destroyer (USA)
DDA	Defence Diversification Agency
DDA	Detroit Diesel Allison (USA)
DDASL	Dowty Defence & Air Systems Ltd (UK)
DDC	Data Device Corp. (USA)
DDefS	Director of Defence Studies
DDESB	Department of Defense Explosives Safety Board (USA)
DDG	Guided Missile Destroyer (USA)
DDI	Developing Defence Industry
DDI	Digital Display Indicator
DDN	Defense Data Network (USA)
DDN	Digital Data Network
DDP	Digital Data Processor
DDRE	Danish Defense Research Establishment
DDR&E	Director of Defense Research & Engineering (USA)

DDS	Department of Defense Support (USA)
DDS	Digital Data Set (Canada)
DDT	Digital Data Transmission
DDTD	Digital Data Transmission Device
DDU	Data Distribution Unit
DDU	Digital Display Unit
DDU	Driver Display Unit for land navigation system
DDVR	Displayed Data Video Recorder
DEC	Digital Electronic Control
DEC	Digital Equipment Corp. (Switzerland)
DECA	Defense & Economic Co-operation Agreement (USA)
DECA	Digital Electronic Control Assembly
DECM	Deceptive Electronic Countermeasures
DECS	Digital Engine Control System
DECU	Digital Engine Control Unit
DED	Data Entry Display
DED	Docking & Essential Defects (UK)
DEDAS	Decontamination Emulsion Direct Application System
DEDS	Digital Enhancement Database System
DEE	Digital Encryption Equipment
DEEC	Digital Electronic Engine Control
DEEP	Dangerous Environment Electrical Protection system
DEFA	Direction des Etudes et Fabrications d'Armement (France)
DEFCOMNON	Defense Command, North Norway
DEFCOMNOR	Defense Command, Norway
DEFCOMSONOR	Defense Command, South Norway
DEFCON	Defense Condition. National defense alert status (USA)
DEFCS	Digital Electronic Flight Control System
DEFPA	Defence of Ports & Anchorages (NATO)
DEF.STAN	Defence Standard
DEI	Direction de l'Electronique et de l'Informatique (France)
DEI	Dynamic Engineering Inc. (USA)
DELEX	Destroyer Life Extension programme
DEM	Demolition
DEMON	Demodulated Noise (sonar processing)
DEP	BV Delft Electronische Producten (Netherlands)
DEPC	Defence Expenditure Procurement Committee (NATO)
DERP	Disposable Eye Respiratory Protection
DES	Data Encryption Standard
DES	Defence Engineering Service
DES	Design Environmental Simulator
DES	Dismounted Extension Switch
DESC	Defense Electronic Support center (USA)
DESO	Defence Export Services Organization (UK)
Det	Detachment
DET	Detonate
Det-Tronics	Detector Electronics Corp. (USA)
DEU	Display Electronics Unit
DEW	Directed Energy Weapon
DEW	Distant Early Warning (USA)

DEWIS	Distant Early Warning Identification Zone
DEWS	Digital EW Simulator
DF	Defensive Fire
DF	Direction Finding
DF	Dong Feng. East Wind, Chinese ballistic missile designator
DF	Methylphosphonic difluoride
DFAD	Digital Feature Analysis Data
DFCS	Digital Fire Control System
DFDAU	Digital Flight Data Acquisition Unit
DFDR	Digital Flight Data Recorder
DFGC	Digital Flight Guidance Computer
DFLP	Democratic Front for the Liberation of Palestine
DFR	Direct Fire Rocket (Pakistan)
DFRF	Dryden Flight Research Facility (USA)
DFSC	Defense Fuel Supply Center (USA)
DFSV	Direct Fire Support Vehicle
DFTS	Defence Fixed Telecommunications System (NATO)
DFU	Deployable Flotation Unit
DFVLR	German Aerospace Research Establishment
DFWES	Direct Fire Weapons Effects Simulator
DF/GA	Day Fighter / Ground Attack
DF/SSL	Direction Finding / Single Station Location
DG	Dayton-Granger Inc. (USA)
DG	Directional Gyro
DG	Director General (UK)
DGA	Delegation Generale pour l'Armement (France)
DGA	Dispersed Ground Alert
DGA	Displacement Gyro Assembly
DGAC	Direction Generale a l'Aviation Civile (France)
DGAM	Direccion General de Aramamento y Material (Spain)
DGDA	Director General of Defence Accounts
DGGN	Direction Generale de la Gendarmerie Nationale (France)
DGI	Directional Gyro Indicator
DGITS	Director General Information Technology System (UK)
DGP	Director General of Police (India)
DGSS	Director General Surface Ships (UK)
DGSW (N)	Director General Surface Weapons (Naval) (NATO)
DGU	Display Generator Unit
DGUW	Director General Underwater Weapons (UK)
DGZ	Desired Ground Zero
DHS	Data Handling System
DHSS	Data Handling Sub-System
DI	Developed Item
DI	Direction Indicator
DI Ind	Defence Intelligence Industries (UK)
DIA	Defense Intelligence Agency (USA)
DIA	Direction de l'Infrastructure Air (France)
DIANE	Detection Identification Analyse des Nouveaux Emetteurs. Helicopter threat warning system (France)
DIANE	Digital Integrated Attack Navigation Equipment

DIAS	Distributed Information Architecture for Ships
DIBUA	Defence In Built-Up Areas
DIC	Defence Intelligence Centre (NATO)
DIC	Defense Intelligence College (USA)
DIC of WA	Defence Industries Council of Western Australia
DICASS	Directional Command Active Sonobuoy System
DICU	Display Interface Control Unit
DID	Defence In Depth
DID	Defence Industry Directorate
DIDA	Defence Industry Development & Support Administration (Turkey)
DIFAR	Direction Frequency And Ranging
DIFFTECH	Differential Technology Ltd (UK)
DILOG	Distributed Logic Corp. (USA)
DIMPS	Distributed Message Processing System
DIMTIS	Director-Mounted Thermal Imaging System (UK)
DIN	Deutsche Institut fur Normung (Germany)
DINAS	Digital Inertial Navigation / Attack System
DINS	Digital Inertial Navigation System
DIP	Digital Image Processing
DIPS	Defense Instantanee Position Strategique (France)
DIPS	Dynamic Isotope Power Subsystems
DIR	Digital Instrumentation Range
DIRAMS	Dockyard Installed Reactor Accident Monitoring System
DIRCEN	Direction des Centres d'Experimentations Nucleaires (France)
DIRSUP	Direct Intelligence Support (NATO)
DIS	Defence Intelligence Staff (NATO)
DIS	Defense Investigative Service (USA)
DIS	Distributed Intelligence System
DISA	Dansk Industri Syndikat A/S (Denmark)
DISA	Defense Information Systems Agency (USA)
DISC	Defense Industrial Support Center (USA)
DISCOM	Division Support Command (USA)
DISCON	Distributed Integrated Secure Communications Network (Australia)
DISTAFF	Directing Staff (Exercise)
DISTAN	Distributed Interactive Secure Telecoms (USA)
DITACS	Digital Tactical System
DITCC	Defence Information Technology Co-ordinating Committee
DITMB	Defence Information Technology Management Board
DITS	Digital Information Transfer System
DIU	Data Interface Unit
DIU	Digital Interface Unit
Div	Division
DIV	Data Over Voice
DIVAD	Division Air Defence
DIVADS	Division Air Defense Gun System (USA)
DJE	Deception Jamming Equipment
DKP	Decontamination Kit, Personal (UK)
DL	Deputy Lieutenant
DLA	Defense Logistics Agency (USA)
DLA	Direction de Lancement d'Armes. Weapon fire control system (France)

DLC	Direct Lift Control
DLC	Downlink Communications
DLCRJ	Detect, Locate, Classify, Record & Jam
DLG	Large Destroyer, Guided Missile
DLMS	Digital Land Mass System
DLOS	Dismountable Line-Of-Sight. Communications
DLOS	Disturbed Line-Of-Sight
DLPP	Data Link Pre-Processor
DLPS	Data Links Processor System
DLR	Deutsche Forschungsanstalt fur Luft-u. Raumfahrt (Germany)
DLS	Data Loader System
DLSA	Director, Land Service Ammunition (UK)
DLSC	Defense Logistics Services Center (USA)
DLSSD	Defense Logistics System Standardisation Division
DLT	Data Link Terminal
DLT	Direction de Lancement de Torpilles. Torpedo fire control system (France)
DLTU	Dismounted Line Termination Unit
DM	Adamsite. Chemical warfare vomiting agent
DM	Data Management
DM	Dorne & Margolin Inc. (USA)
DMA	Defense Mapping Agency (USA)
DMA	Delay Arming Mechanism
DMA	Direct Memory Access
DMA	The Defence Manufacturers Association of Great Britain Ltd (UK)
DMAA	The Defence Manufacturers Association of Australia Limited
DMAB	Defended Modular Array Basing
DMB	Dvadesetprvi Maj Beograd Ro TMT (Yugoslavia)
DmC	Datametrics Corp. (USA)
DMC	Daniels Manufacturing Corp. (USA)
DMC	Defence Manufacturers Council (Australia)
DMD	Digital Message Device
DME	Distance Measuring Equipment
DME-P	Distance Measuring Equipment Precision
DMG	Digital Map Generator
DMIP	DCS Mediterranean Improvement Program
DML	Decision & Modelling Language
DML	Devonport Management Ltd (UK)
DMO	Directorate of Military Operations (UK)
DMR	Digital Microwave Radio
DMS	Data Monitoring System
DMS	Data Multiplexer Sub-unit
DMS	Digital Mapping System
DMS	Direct Moulded Sole
DMS	Domestic Military Sales
DMS	Dual Mode Seeker
DMSK	Differential Minimum Shift Keying
DMSP	Defence Meteorological Satellite Programme
DMSP	Dual Mode Speech Processor

DMT	Deep Mobile Target
DMT	Dual Mode laser / TV Tracker
DMTI	Digital Moving Target Indicator
DMTS	Deck-Mounted Torpedo-launch System
DMU	Data Management Unit
DMU	Digital Master Unit
DMVS	Desert Mobility Vehicle System
DMZ	Demilitarised Zone
DNA	Defense Nuclear Agency (USA)
DNBCC	Defence NBC Centre (UK)
DNCCC	Defense National Communications Control Center (USA)
DND	Department of National Defence (Canada)
DNI	Director of National Intelligence
DNO	Directorate Naval Operations
DNRS	Day / Night Range Sight
DNS	Direct Network Subscriber
DNST	Directorate Naval Shore Telecommunications
DNSy	Director of Naval Security
DnV	Det norske Veritas (Norway)
DNVT	Digital Non-secure Voice Terminal
DO	Divisional Officer (Royal Navy)
DOA	Direction Of Arrival
DOAE	Defence Operational Analysis Establishment (UK)
DOC	Delayed Opening Chaff
DOC	District Officer Commanding
DOCC	DCA Operational Control Complex (USA)
DOCC	Director of the Operational Command Committee (Brunei)
DoD	Department of Defense (USA)
DOD MILSO	Department Of Defense Military Standard Logistics Systems Office (USA)
DODCI	Department Of Defense Computer Institute (USA)
DODIIS	Department of Defense Intelligence Information System (USA)
DoE	Department of Energy
DOLE	Detection Of Laser Emissions
DOLRAM	Detection Of Laser, Radar & Millimetric Threats
DOM TOM	Departments et Territories d'Outre Mer (France)
DOMS	Defence Operational Movement Staff
DOP	Department Of Productivity
DOP	Drop-Off Point
DOPIN	Dispenser / On-board Processor & Inertial Navigation
DOPT	Defence Organisation Project Team (UK)
DORNA	Direccion de tiro Optronica y Radarica Naval. Combined radar & electro-optical fire control system (Spain)
DORSI	Dornier Radar Scoring Indication System
DOT	Designating Optical Tracker (now known as Queen Match)
DOV	Discreet Operational Vehicle
DOW	Digital Orderwire channel
DP	Data Processing
DP	Diphosgene. Chemical warfare choking agent
DP	Dual Purpose (gun for surface or AA use)

DPC	Defence Planning Committee (NATO)
DPCM	Differential Pulse Code Modulation
DPCM	Digital Pulse Code Modulation
DPEE	Directorate of Proof & Experimental Establishments (UK)
DPG	Data Processor Group
DPICM	Dual Purpose Improved Conventional Munition (USA)
DPM	Defense Products Marketing Inc. (USA)
DPM	Digital Plotter Map
DPM	Disruptive Pattern Material
DPO	Defence Press Office (UK)
DPP	Defence Planning & Policy (NATO)
DPQ	Defence Planning Questionnaire (NATO)
DPR-RI	Dewan Perwakilan Rakyat Republik Indonesia
DPSK	Digital Phase Shift Keying
DPSN	Defence Packet Switched Network
DPU	Data Processing Unit
DPU	Digital Processing Unit
DPU	Display & Processing Unit
DQ	Data Qualifier
DQA/TS	Directorate of Quality Assurance / Technical Support
DR	Dead Reckoning
DRA	Defence Research Agency (UK)
DRACAS	Data Reporting, Analysis & Corrective Action System
DRAMA	Digital Radio And Multiplex Acquisition (USA)
DRASH	Deployable Rapid Assembly Shelter
DRB	Defense Resources Board (USA)
DRC	Defence Review Committee (NATO)
DRDL	Defence Research & Development Laboratories (India)
DRDO	Defence & Research Development Organization (India)
DREA	Defence Research Establishment Atlantic (Canada)
DREP	Defence Research Establishment Pacific (Canada)
DRES	Defence Research Establishment Suffield (Canada)
DRET	Direction des Recherches, Etudes de Techniques (France)
DRF	Dual Role Fighter
DRFM	Digital Radio Frequency Memory
DRG	Defence Research Group (NATO)
DRI	Delegation aux Relations Internationales (France)
DRIC	Defense Research Information Center (USA)
DRIR	Direct Readout Infra-Red
DRISDE	Directorate General of Information & Social Relations of Defense (Spain)
DRIU	Damage Repair Instruction Unit
DRLM	Digital Radar Land Mass
DRLMS	Digital Radar Land Mass Simulator
DRM	Digital Range Meter
DRMR	Defense Re-utilization & Marketing Region (USA)
DRMR	Defense Reutilization & Marketing Region-Ogden (USA)
DROPS	Demountable Rack Off-loading & Pick-up System
DRPS	Defence Radiological Protection Service (UK)
DRS	Data Relay Satellite

DRS	Detection & Ranging Set (USA)
DRS	Diagnostic/Retrieval Systems Inc. (USA)
DRT	Defense Research Technologies Inc. (USA)
DRTP	Dynatech Real Time Products (UK)
DRTS	Detecting, Ranging & Tracking System
DRU	Data Retrieval Unit
DRU	Drive Unit
DRU	Dynamic Reference Unit
DS	Direct Support
DS	Directionally Solidified
DS	Discarding Sabot
DSA	Defence Support Agency
DSAA	Defense Security Assistance Agency (USA)
DSABS	Defence Science Advisory Board (Canada)
DSACEUR	Deputy Supreme Allied Commander, Europe (NATO)
DSACS	Direct Support Armored Cannon System (USA)
DSAR	Defense Supply Agency Regulation
DSARC	Defense Systems Acquisition Review Council (USA)
DSB	Defence Science Board
DSB	Deutsche Schlauchbootfabrik Hans Scheibert GmbH & Co. KG (Germany)
DSB	Double Side Band
DSB	Drug Supervisory Body
DSB	Duracell Special Batteries (UK)
DSC	Deputy Squadron Commander (Royal Air Force)
DSC	Digital Scan Convertor
DSC	Direction de la Securite Civile (France)
DSCS	Defense Satellite Communications System (USA)
DSD	Data Systems Designers (Israel)
DSD	Defence Support Division (NATO)
DSETS	Direct Support Electrical Test System
DSIC	Dowty & Smiths Industries Controls (UK)
DSIS	Digital Software Integration System
DSMAC	Digital Scene Matching Area Correlation
DSN	Defense Switched Network (USA)
DSO	Defence Sales Organisation
DSO in C	Deputy Signal Officer-in-Chief (UK)
DSP	Defence Support Program (USA)
DSP	Digital Signal Processor
DSPG	Defense Special Projects Group (USA)
DSPMO	SAMMS Program Management Office (USA)
DSREDS	Digital Storage & Retrieval Engineering Data System
DSRV	Deep Submergence Rescue Vehicle (USA)
DSS	Data Storage Set
DSSO	Defense Systems Support Organization (USA)
DSSS	Direct Sequence Spread Spectrum
DST	Defence Suppression Threat
DST	Defence & Space Talks
DSTO	Defence Science & Technology Organisation (Australia)
DSU	Data Storage Unit

DSU	Direct Support Unit (USAF)
DSUR	Data Storage Unit Receptacle
DSV	Deep Submergence Vehicle (USA)
DSV	Defence Suppression Vehicle
DSVT	Digital Secure Voice Terminal
DSVT	Digital Subscriber Voice Terminal
DSWP(N)	Director, Surface Weapons Projects (Naval) (UK)
DSWS	Division Support Weapon System
DS/T	Discarding Sabot / Tracer
DT	Data Transmission
DTAT	Direction Technique des Armements Terrtestres (France)
DTC	Data Transfer Cartridge
DTC	Dynatech Tactical Communications (USA)
DTCN	Direction Techniques Constructions Navales (France)
DTCS	Drone Tracking & Control System
DTD	Data Transfer Device
DTE	Data Terminal Equipment
DTE	Data Transfer Equipment
DTED	Digital Terrain Elevation Data
DTG	Digital Transmission Group
DTG	Distance To Go
DTG	Dynamically Turned Gyro
DTH	Down-The-Hill radios
DTI	Department of Trade & Industry (UK)
DTI	Digital Technology Inc. (USA)
DTIC	Defense Technical Information Center (USA)
DTM	Data Transfer Module
DTM-D	Digital Terrain Management & Display
DTMF	Dual-Tone Multiple-Frequency
DTN	Data Transmission Network
DTO	Data Take-Off unit
DTOC	Divisional Tactical Operations Centre
DTS	Data Terminal Set
DTS	Data Transfer System
DTS	Defense Transportation System
DTS	Desarrollo de Tecnologias y Sistemas Ltda (Chile)
DTST	Defence Technology Study Team
DTTS	Dynamic Track Tensioning System
DTU	Data Terminal Unit
DTU	Data Transfer Unit
DTU	Data Transmission Unit
DTU	Distance Transmitter Unit
DT&E	Development, Test & Evaluation
DU	Depleted Uranium
DU	Detector Unit
DU	Display Unit
DUAD	Dual Air Defence
DUI	Diving Unlimited International Inc. (USA)
DUS	Datchik Uglovykh Skorostej. Angular speed sensor (Romania)
DUWIR	Dual Waveband Imaging Radiometer (UK)

DVI	Direct Voice Input
DVITS	Digital Video Imaging Transmission System
DVL	Data / Voice Logger
DVOR	Doppler VHF Omni-directional Ranging Equipment
DVRI	Direct View Radar Indicator
DVRS	Display Video Recording System
DVT	Delta Voice Terminal
DVVD	Dual Vision Viewing Device
DVX	Digital Voice Exchange
DV/A	Doppler Velocimeter / Altimeter
DWFK	Deep Water Fording Kit
DWI	Direct Wireless Installation for exploding mines
DWP	Defence White Paper
DWRNS	Director, Womens Royal Naval Service (UK)
DWS	Dispenser Weapon System
DWT	Deutsche Gesellschaft fur Wehrtechnik (Germany)
DWTS	Digital Wideband Transmission System
DZ	Dropping Zone
D&V	Demonstration & Validation programme
D(A)AG	Deputy (Assistant) Adjutant-General
D/CINCENT	Deputy Commander, Allied Forces Central Europe (NATO)
D/CM	Diagnostic & Condition Monitoring

E

E in C	Engineer-in-Chief (UK)
E2I	Exoatmospheric-Endoatmospheric Interceptor
E3	End-to-End Encryption
EA	Elettronica Aster Spa (Italy)
EA	Enemy Aircraft
EA	Engineering Authority
EAA	Electro-Acoustic Subassembly
EAA	Experimental Aircraft Association
EAA	External Access Applique
EAAK	Enhanced Applique Armor Kit (USA)
EAB	External Access Box
EAC	Embraer Aircraft Corp. (USA)
EACOS NMR	Executive Assistant to the Chief of Staff for National Military representatives, SHAPE (NATO)
EADI	Electronic Attitude Director Indicator
EAD/EAC	Echelon Above Division / Echelon Above Corps
EA-EMA	Ecole de l'Air et Ecole Militaire de l'Air (France)
EAM	Emergency Action Message
EAOS	Enhanced Artillery Observer Sub-system
EAP	Experimental Aircraft Programme
EAR	Electronically Agile Radar
EAROM	Electrically Alterable Read-Only Memory
EAS	Equivalent Airspeed
EASI	Engineered Air Systems Inc. (USA)
EASTLANT	Eastern Atlantic Area (NATO)
EAT	Estimated Approach Time
EAT	European Advanced Technologies (Belgium)
EATC	European Aviation Training Center NV (Belgium)
EAU	Emergency Action Unit (NATO)
EAU	Engine Analyser Unit
EBC	Emergency Beacon Corp. (USA)
EBG	Elektronische Bauelemente Ges.mbH (Austria)
EBG	Engin Blinde de Genie. Combat engineer tractor (France)
EBR	Engin Blinde de Reconnaissance. Armoured reconnaissance vehicle (France)
EBTR	Electronic Bearing Time Recorder
EC	Earth Coverage antenna
EC	Electronic Combat
ECA	Electronic Control Amplifier
ECA	European Combat Aircraft
ECA	Experimental Combat Aircraft
ECAM	Electronic Centralised Aircraft Monitor
ECAN	Etablissement des Constructions et Armes Navales (France)
ECAS	Electronic Chemical Agent Alarm System
ECC	Export Consultants Corp. (USA)
ECCAP	High explosive anti-tank & anti-personnel
ECCD	Electric Cockpit Control Device

ECCM	Electronic Counter-Countermeasures
ECDES	Electronic Combat Digital Evaluation System (USAF)
EC-EC	Earth Coverage to Earth Coverage
ECI	Erich Cordes Industrievertretungen (Germany)
ECIA	Esperanza y Cia SA (Spain)
ECIF	Electronic Components Industry Federation (UK)
ECIP	Energy Conservation Investment Programme
ECL	Emitter Coupled Logic
ECLSS	Environmental Control & Life-Support System
ECM	Electronic Countermeasures
ECMES	Electronic Combat Modelling & Evaluation System
ECMRITS	Electronic Countermeasures Resistant Information Transmission System
EC-NC/AC	Earth Coverage to Narrow Coverage / Area Coverage
ECNI	Enhanced Communications, Navigation & Identification
ECOP	Electronic Co-Pilot
ECOSEN	Equipos de Control y Senalizacion CA (Venezuela)
ECOWAS	Economic Community of West African States
ECR	Electronic Combat & Reconnaissance
ECR	Embedded Computer Resources
ECRIEE	East China Research Institute of Electronic Engineering
ECS	Engagement Control Station
ECU	Electronic Control Unit
ECU	Environmental Control Unit
ECU	Exercise Control Unit
ECW	Electronic Combat Wing
EC/NBC	Environmental Control / Nuclear, Biological & Chemical
ED	Engineering Dynamics (Southern) Ltd (UK)
ED	Ethyldichloroarsine. Chemical warfare blister agent
ED	Explosive Device
EdA	Ejercito del Aire (Spain)
EDA	Electronic Development Associates Inc. (USA)
EDATS	Extra Deep Armed Team Sweep
EDAU	Extended Data Acquisition Unit
EDB	Extruded Double Base
EDC	Electronics Development Corp. (USA)
EDC	Environmental Discrimination Circuit
EDC	Error Detection & Correction
EDC	European Defence Community
EDCARS	Engineering Data Computer Assisted Retrieval System
EDCC	Emergency Defence Communications Centre (UK)
EDICS	European Defence Industry Collaboration Service
EDIG	European Defence Industrial Group
EDIP	European Defence Improvement Programme (NATO)
EDIR	Ecartometrie Differentielle Infra-Rouge. Infra-red missile director (France)
EDIS	European Defence Industry Study
EDL	Electron Discharge Laser
EDM	Electronic Distance Measurer
EDM	Engine Data Multiplexer

EDM	Engineering Development Model
EDMAG	European Defence Manufacturers Group
EDP	Electronic Data Processor
EDS	Electronic Data Systems Corp. (USA)
EDT	Eastern Daylight Time (USA)
EDTS	Expanded Data Transfer System
EDU	Electronic Display Unit
EDU	Engine Diagnostic Unit
EDU	European Democratic Union
EEAD	Extended Air Defence
EEC	Electronic Engine Controls
EEC	European Economic Community
EEC	High explosive, general-purpose
EED	Electro-Explosive Device
EEF	Effective Enemy Fire
EEPROM	Electrically Erasable Programmable Read-Only Memory
EET	Eastern European Time
EET	Emergency Evacuation Trainer
EEWT	Elbit Electronic Warfare Trainer (Israel)
EEZ	Economic Exclusion Zone
EFA	European Fighter Aircraft
EFAB	Etablissement d'Etudes et de Fabrication d'Armement de Bourges (France)
EFC	Equivalent Full Charge
EFCR	Equivalent Full Charge Rounds
EFCS	Electronic Control System
EFCS	Electronic Filmless Camera System (USA)
EFCS	Enhanced Fire Control System
EFCU	Electrical Flight Control Unit
EFDARS	Expansible Flight Data Acquisition & Recording System
EFF	Explosively Formed Fragment
EFIS	Electronic Flight Instrumentation System
EFM	Enhanced Fighter Manoeuvrability
EFM	Explosives Factory Maribyrnong (Australia)
EFP	Expanded Feasibility Phase
EFP	Explosively Formed Penetrator
EFP	Explosively Formed Projectile
EFS	Explosionsunter-druckungssystem. Explosion suppression system (Austria)
EFTA	European Free Trade Association
EFVS	Electronic Fighting Vehicle System
EGES	Ensemble de Gestion et Exploitation Systeme. FSAF command system (France)
EGI	Embedded GPS/INU programme (USA)
EGTD	Electric Gun & Turret Drive
EHA	Electro-Hydrostatic Actuator
EHDD	Electronic Head-Down Display
EHI	EH Industries Ltd (UK)
EHSI	Electronic Horizontal Situation Indicator
EIA	Electronic Industries Association (USA)

EICAS	Engine Indication & Crew Alerting System
EID	Empresa de Investigacao e Desenvolvimento de Electronica SA (Portugal)
EIDEL	Eidsvoll Electronics AS (Norway)
EINSA	Equipos Industriales de Manutencion SA (Spain)
EIRP	Equivalent Isotropically Radiated Power
EIS	Electronic Instrument System
EIS	Entry Into Service
EIS	Environmental Impact Statement (USA)
EIU	Electronic Interface Unit
EJS	Enhanced JTIDS System (USA)
EL	Electro-Luminescent
EL	Emitter locator
ELA	Extender Lens Assembly (USA)
ELAC	Elevator & Aileron Computer
Elbasa	Electronica Basica SA (Spain)
ELBO	Hellenic Vehicle Industry SA (Greece)
ELCAS	Elevated Causeway
ELDO	European Space Launcher Development Organization. Now known as ESA
ELF	Eritrean Liberation Front (Ethiopia)
ELINT	Electronic Intelligence
ELKE	Elevated Kinetic Energy Weapon (USA)
ELPA	Eyring Low-Profile Antenna
ELS	Electro-magnetic Launcher System
ELS	Emitter Location System
ELSA	Electronic Lobe-Switching Antenna
ELSAP	Elektronische Schiessanlage fur Panzer. Electronic fire system for tanks (Germany)
ELSS	Environmental Life Support System
ELT	Emergency Locator Transmitter
ELTA	EHF Lightweight Transportable Antenna
ELV	Expendable Launch Vehicle
ELV	Experimental Launch Vehicle
EM	Electro-Magnetic
EM	Elicotteri Meridionali SpA (Italy)
EM	Engineered Magnetics Inc. (USA)
EMA	Electronic Missile Acquisition
EMA	European Monetary Agreement
EMA	External Mounting Assembly
EMATT	Expendable Mobile ASW Training Target
EMC	Electro-Magnetic Compatibility
EMC	Elektronik Mekanik Cihaziar Ticaret Ltd (Turkey)
EMC	Executive Management Committee
EMCDB	Elastomer Modified Cast Double Base
EMCOF	European Monetary Co-operation Fund
EMD	Engine Model Derivative
EMD	Engineering & Manufacturing Development
EMDG	Euromissile Dynamics Group (France)
EMF	European Monetary Fund

EMFTD	Electro-Magnetic Focused Technology Demonstrator
EMG	Electro-Magnetic Gun
EMG	Externally Mounted Gun
EMI	Electro-Magnetic Interference
EMIH	Electro-Magnetic Interference Hardening
EMIS	Electro-Magnetic Isotope Separation
EML	Electro-Magnetic Launcher
EML	Electronic Microsystems Ltd (Israel)
EMM	Electronic Memories & Magnetics NV (Belgium)
EMMPS	Enhanced MEECN Message Processing System (USA)
EMMU	Engine Monitor Multiplexer Unit
EMP	Electro-Magnetic Pulse
EMP	Engineering Mine Plough
EMPAR	European Multi-function Phased-Array Radar
EMPASS	Electro-Magnetic Performance of Air & Ship System
EMPP	Electro-Magnetic Pulse Protection
EMR	Electro-Magnetic Radiation
EMRLD	Excimer Repetitively Pulsed Laser Device
EMRU	Electro-Magnetic Release Unit
EMS	Electro Mechanical Systems Ltd (UK)
EMS	Electro-Magnetic pulse Shielding
EMS	Enhanced Mobility System
EMS	Equipment Maintenance Squadron (USA)
EMS	European Monetary System
EMSC	Engine Monitoring System Computer
EMTE	Electro-magnetic Test Environment (USA)
EMU	Electromagnetic Unit
EMU	Electronic Message Unit
EMU	Elevation Measuring Unit
EMux	Electronic Multiplexing
ENAER	Empresa Nacional de Aeronautica (Chile)
ENDEX	End of Exercise (UK)
Eng	Engineering
ENGESA	Engesa Engenheiros Especializados (Brazil)
ENK	Endo-atmospheric Non-nuclear Kill
ENOSA	Empresa Nacional de Optica SA (Spain)
Ens	Ensign
ENSCE	Enemy Situation Correlation Element (USA)
ENSIP	Engine Structural Integrity Program (USAF)
EO	Electro-Optic
EOB	Electronic Order of Battle
EOCCM	Electro-Optical Counter-Countermeasures
EOCM	Electro-Optical Countermeasures
EOD	Electro Optic Developments Ltd (UK)
EOD	Electro-Optic Device
EOD	Explosive Ordnance Demolition
EOD	Explosive Ordnance Disposal
EOFC	Electro-Optical Fire Control
EOG	Elaboratuer d'Ordre de Guidage. Guidance command processor (France)

EOIS	Electro-Optical Imaging System
EOR	Explosive Ordnance Reconnaissance
EORSAT	Electronic Ocean Reconnaissance Satellite
EOSS	Electro-Optic Sensor System
EOW	Engineering Order Wire circuits
EP	Exercise Practice
EPA	Environmental Protection Agency
EPC	Electro Prismatic Collimator
EPC	Electronic Plane Conversion
EPC	Engin Principal de Combat. Future main battle tank (France)
EPCI	Enhanced Proliferation Control Initiative (USA)
EPG	European Programme Group
EPG	Externally Powered Machine Gun
EPI	Engine Performance Indicator
EPI	TRW Electronic Products Inc. (USA)
EPIN	Ente Promozione Industria per la Difesa Navale (Italy)
EPLF	Eritrean People's Liberation Front (Ethiopia)
EPLRS	Enhanced Position Location Reporting System
EPMaRV	Earth Penetrating Manoeuvring Re-entry Vehicle
EPNdB	Effective Perceived Noise Decibel
EPNL	Effective Perceived Noise Level
EPO	Evershed Power Optics
EPP	Electric Power Plant
EPPU	Electronic Pack Processor Unit
EPR	Engine Pressure Ratio
EPRDU	Electronic Pack Remote Display Unit
EPREP	Explosives, Propellants & Related End Products agreement (UK)
EPROM	Erasable Programmable Read Only Memory
EPS	Engineering & Professional Services Inc. (USA)
EPU	Electrical Power Unit
EPU	Emergency Power Unit
EPUU	Enhanced PLRS User Unit. Communications
ER	Enhanced Radiation
ER	Established Reliability
ER	Extended Range
ERA	Engineering Research Associates Inc. (USA)
ERA	Explosive Reactive Armour
ERA	Extended Range Ammunition
ERADCOM	Electronics Research & Development Command
ERAM	Extended Range Anti-armor Munition
ERAM	Extended Range Anti-tank Mine
ERAPDS	Enhanced Recognised Air Picture Dissemination System
ERAPIDS	Extended Recognised Air Picture Display System (NATO)
ERAPS	Expendable Reliable Acoustic Path Sonobuoy
ER-ATACMS	Extended-Range Army Tactical Missile System (US Army)
ERATO	Extended Range Automatic Targeting of Otomat missile (France)
ERC	Earth Resources Consultants (USA)
ERC	Engin de Reconnaissance Canon (France)
ERCO	Engineering Representation & Consultants Co. (Egypt)
ERCS	Emergency Rocket Communications System (USA)

ERDE	Explosives Research & Development Establishment
E-RETS	Enhanced Remote Target System
ERFB	Extended Range Full Bore
ERFB-BB	Extended Range Full Bore, Base Bleed
ERGFCDS	Extended Range Gunnery Fire Control Demonstration System (USA)
ERGP	Extended Range Guided Projectile
ERINT	Extended Range Interceptor
ERIS	Exoatmospheric Re-entry vehicle Interception System
ERMIS	Extended Range Modification Integration System
EROM	Erasable Read-Only Memory
ERP	Effective Radiated Power
ERP	Extended Range Projectile
ERP	Extended Range Proximity
ERRB	Enhanced Radiation Reduced Blast
ERS	European Remote sensing Satellite
ERSC	Extended Range Sub-Calibre
ERU	Ejector Release Unit
ERU	Emergency Reaction Unit
ERV	Electronic Repair Vehicle
ERV	Emergency Rendezvous
ERV	Emergency Rescue Vehicle
ERVIS	Exoatmospheric Re-entry Vehicle Interception System (USA)
ERW	Enhanced Radiation Weapon
ERWE	Enhanced Radar Warning Equipment
ER/DL	Extended Range Data Link
ER/RB	Enhanced Radiation / Reduced Blast
ESA	Electronic Signature Authentication
ESA	European Space Agency
ESA	Office of Engineering Support Activity, Naval Weapons (USA)
ESAS	Electronically Steerable Antenna System
ESC	Engine Supervisory Control
ESC	European Security Community
ESCAN	Electronics & Space Corporation of Canada Ltd
ESCO	Engineered Systems Co. (USA)
ESD	Electronic Systems Division (USAF)
ESD	Electronique Serge Dassault. Now Dassault Electronique (France)
ESE	Electronic Security Environment
ESG	Electrostatically Suspended Gyro
ESG	Elektronik-System-Gesellschaft mbH (Germany)
ESM	Electronic Support / Surveillance Measures
ESMACO	Estado Mayor Conjunto (Uruguay)
ESOC	European Space Operations Centre (Germany)
ESP	Expendable Signal Processor
ESP	Expendables System Programmer
ESPA	Equipements Speciaux pour l'Aviation (France)
ESPAWS	Enhanced Self-Propelled Artillery Weapon System (USA)
ESPRIT	Eye-Slaved Projected Raster Inset
ESR	Electro-Slag Refined steel
ESRO	European Space Research Organization (now ESA)
ESS	Exercise Support System

ESS	Exploitation Support Segment (JSIPS)
ESS	External Suspension System
ESSCO	Electronic Space Systems Corp. (USA)
ES-SDMS	Expanded Service-Shipboard Data Multiplex System
ESSI	Electronic Support Systems Inc. (USA)
ESSM	Evolved Seasparrow Missile
EST	Eastern Standard Time
EST	Electronic Support & Training
ESTC	Explosives Storage & Transport Committee (UK)
ESTEC	European Space Research & Technology Centre (Netherlands)
ESU	Electronic Storage Unit
ESU	External Stabilisation Unit
ET	Eastern Time
ET	Electronic Timer
ET	Electrothermal
ET	Emerging Technology
ETA	Estimated Time of Acquisition
ETA	Estimated Time of Arrival
ETAS	Etablissement Technique d'Angers (France)
ETBS	Etablissement Technique de Bourges (France)
ETC	Electrothermal-Chemical
ETC	Environmental Tectonics Corp. (USA)
ETD	Estimated Time of Departure
ETE	Estimated Time Enroute
ETEC-E	Electronics & Telecommunications Evaluation Center – Europe (Germany)
ETEM	Empresa de Technologia e Equipamento Militar Lda (Portugal)
ETES	Exotic Threat Emitter System
ETF	Electronic Time Fuze
ETF	Enhanced Tactical Fighter
ETMP	Enhanced Terrain Masked Penetration
ETP	Estimated Time of Penetration
ETPU	Engine Transient Pressure Unit
ETR	Electric Target Range
ETR	Engineered To Reliance
ETS	European Telephone System
ETSS	Electronic Telecommunications Switching System (USA)
ETS/N	Engagement Training Simulator Network
ET&T	European Test & Evaluation (USAF)
EUB	Essential User Bypass. Communications
EUCLID	European Co-operative Long-Term Initiative in Defence
EUCOM	European Command (NATO)
EURATOM	European Atomic Energy Committee
Euro-ART	European Advanced Radar Technology
EUROCHEMIC	European Company for Chemical Processing of Iradiated Fuels
EUROCOM	European Communications
EUROCONTROL	European Organization for the Safety of Air Navigation
EUROFER	European Steel Federation
EURONET	European Information Network

EUROSPACE	European Industrial Space Study Group
EUV	Extreme Ultra-Violet
EVA	Extra-Vehicular Activity
EVER	Endurance Vehicle for Extended Reconnaissance
EW	Early Warning
EW	Electronic Warfare
EWACS	Electronic Warfare Analysis Centre
EWAISF	Electronic Warfare Avionics Integrated Support Facility
EWAM	Extended Window Addressable Memory. Large scale integrated circuit
EWAS	Electronic Warfare Analysis System
EWAU	Electronic Warfare Avionics Unit (Royal Air Force)
EWC	The Electric Wire & Cable Co. of Israel Ltd
EWEDS	Electronic Warfare Evaluation Display System
EWEP	Electronic Warfare Evaluation Program (USA)
EWES	Electronic Warfare Evaluation System
EWG	Executive Working Group (NATO)
EWK	Eisenwerke Kaiserslautern (Germany)
EWO	Electronic Warfare Officer
EWO	Emergency War Order
EWOC	Electronic Warfare Operator's Console
EWOSE	Electronic Warfare Operational Support Establishment (UK)
EWS	External Weapon Station
EWSM	Early-Warning Support Measures
EWT	Engineer Wheeled Tractor
EWTS	Electronic Warfare Training System
EWTS-R	Electronic Warfare Training System – Radar
EW&C	Early Warning & Control
EW/GCI	Early Warning & Ground-Controlled Intercept
EX	Electronics Experimental (USA)
EXCM	External Countermeasures
ExE	Event by Event
EXJAM	Expendable Jamming System
EZP	Exclusion Zone Patrol
E&E	Escape & Evasion

F

FA	Soviet Frontal Aviation
FAA	Federal Aviation Administration (USA)
FAA	Fleet Air Arm (UK)
FAA	Fuerza Aerea Argentina
FAACA	Force Anti-Air Co-ordination Area
FAAD	Forward Area Air Defense (USA)
FAADC2I	Forward Area Air Defence Command, Control & Intelligence
FAADS	Forward Area Air Defense System (USA)
FAAG	Commandement des Forces Aeriennes aux Antilles-Guyane (France)
FAAITM	Fleet Air Arm Information Technology Management (UK)
FAALS	Forward Area Armored Logistic System
FAAM	Family of Air-to-Air Missiles
FAAR	Forward Area Alerting Radar (USA)
FAASV	Field Artillery Ammunition Support Vehicle (USA)
FAB	Forca Aerea Brasileira. Brazilian Air Force
FAB	Fugasnaya Aviabomba. General-purpose bomb (CIS)
FABA	Fabrica de Artilleria de Bazan (Spain)
FAC	Fast Attack Craft
FAC	Flight Augmentation Computer
FAC	Forward Air Controller
FACA	Future Attack & Combat Aircraft
FACE	Field Artillery Computing Equipment (UK)
FACON	Facilities Control
FACP	Forward Air Control Post
FACTS	FLIR Augmented Cobra TOW Sight
FAD	Fleet Air Defense (USA)
FADAC	Field Artillery Digital Automatic Computer
FADEC	Full Authority Digital Engine Control
FAE	Fuel-Air Explosive
FAE	Fuerza Aerea Ecuatoriana (Ecuador)
FAESHED	Fuel-Air Explosive Helicopter Delivered
FAFC	Full Authority Fuel Control
FAI	Federation Aeronautique Internationale (France)
FALCON	Fission Activated Laser Concept
FALCON	Frequency-Agile Low Coverage Netted radar
FALCON	Fuel / Air Line Charge Ordnance Neutraliser
FALW	Family of Air-Launched Weapons
FAM	Field Alarm Module
FAM	Fighter Attack Manoeuvring
FAM	Flexible Attrition Model. Simulation system
FAME	Frequency Assignment Management Equipment
FAMET	Fuerzas Aeromoviles del Ejercito de Tierra (Spain)
FAMG	Field Artillery Missile Group (USA)
FAMS	Family of Anti-air Missile Systems
FAMS	Field Automatic Message Switch
FAMV	Forward Area Multi-purpose Vehicle
FAN	Force d'Action Navale (France)

FAN	Forces Armees du Nord. Northern Armed Forces (Chad)
FAN	Forward Air Navigator
Fanazul	Fabrica Militar Polvoras y Explosivos "Azul" (Argentina)
FANCI	Forces Armees Nationales de la Cote D'Ivoire (Ivory Coast)
FANY	First Aid Nursing Yeomanry
FAP	Final Approach
FAPA	Force Area Populaire de Angola
FAPDS	Frangible Armour-Piercing, Discarding Sabot
FAPLA	Angolan People's Liberation Army
FAR	False Alarm Rate
FAR	Federal Acquisition Regulation (USA)
FAR	Force d'Action Rapide (France)
FARCOS	Fast Adaptive HF Radio Communication System
FARE	Forward Area Refuelling Equipment
FARP	Forward Area Rearm-refuel Point
FARP	People's Revolutionary Armed Forces (Guinea Bissau)
FARRP	Forward-Area Rearming & Refuelling Point
FARS	Fast Acquisition Receiver System
FARS	Field Artillery Rocket System
FARS	Frequency Array Radar System
FARV-A	Future Armored Resupply Vehicle, Artillery (USA)
FAS	Flank Array Sonar (Germany)
FAS	Force Aerienne Strategique (France)
FAS	Forward Acquisition Sensor
FAS	Frequency-Agile Subsystem
FASCAM	Family of Scatterable Mines
FAST	Fly Away Satellite Terminal
FAST	Forward Area Support Team
FAST	Fully Automatic Scoring Target
FAST	Fuze Activating Static Targets
FASTA	Flugzeugabwehrstartanlage. Air defence launcher assembly (Germany)
FASTAR	Forward Area Surveillance & Target Acquisition System
FASTC	Foreign Aerospace Science & Technology Directorate Center (USAF)
FASTHAT	Fast High-Accuracy Tunable receiver
FASTNET	Fixed Army Strategic Telephone Network
FAT	First Article Testing
FATAC	Commandement des Forces Aerienne Tactiques (France)
FATE	Fuzing, Arming, Test & Evaluation
FATG	Fixed Air-To-Ground
FATS	Field Automatic Telephone Switch
FATS	Firearms Training System
FATS	Firearms Training Systems Inc. (USA)
FATT	Forward Area Tactical Teletype (USA)
FAUSA	Fokker Aircraft USA Inc.
FAV	Fast Attack Vehicle (USA)
FAW	Fighter, All-Weather (UK)
FAWPSS	Forward Area Water Point Supply System
FAX	Facsimile
FB	Fighter, Bomber (UK)

FBI	Federal Bureau of Investigation (USA)
FBL	Fly-By-Light
FBM	Fleet Ballistic Missile
FBS	Fly-By-Speech
FBW	Fly-By-Wire
FC	Corvette
FC	Fire Control
FCC	Fire Control Centre
FCC	Fleet Command Centre
FCC	Flight Control Computer
FCCT	Field Communication Centre Terminal
FCCVS	Future Close-Combat Vehicle System (USA)
FCDC	Flight Control Data Recorder
FCE	Fire Control Equipment
FCEU	Fire Control Electronics Unit
FCMC	Fire-Control & Monitoring Computer
FCNP	Fire Control Navigation Panel
FCO	Fire Control Officer
FCPEM	Fire Control Processor Electronics Module
FCR	Fire Control Radar
FCS	Field-expedient Mineclearing System
FCS	Fire Control System
FCSS	Fire Control Sight System
FCT	Foreign Comparative Test programme
FCTS	Flight Controller Training System
FC/NP	Fire-Control / Navigation Panel
FD	Fill Device. Communications
FD	Full Development
FDAS	Flight Data Acquisition System
FDAU	Flight Data Acquisition Unit
FDC	Fire Distribution Centre
FDCV	Fire Direction Centre Vehicle
FDDI	Fibre Distributed Data Interface
FDDS	Flag Data Display Systems
FDEP	Flight Data Entry Panel
FDFF	Foreningen af Danske Fabrikanter af Flymateriel (Denmark)
FDI	Federation of Danish Industries
FDIU	Flight Data Interface Unit
FDL	Fast Deployment Logistic
FDM	Fokker Defence Marketing (Netherlands)
FDM	Frequency Division Multiplex
FDMA	Frequency-Division Multiple Access
FDOT	Furezas de Defensa Operativa del Territorio (Spain)
FDS	Fixed Distributed System
FDS	Flight Director System
FDS	Functionally Distributed Simulation
FDU	Flight Data Unit
FE	Frequency Electronics Inc. (USA)
FEAF	Far East Air Force
FEATS	Feasibility & Experimentation in Aquisition & Tracking Systems

FEBA	Forward Edge of the Battle Area
FEC	Forward Error Correction
FEFA	Future European Fighter Aircraft project
FEI	Flightline Electronics Inc. (USA)
FEL	Fibre Elastomeric (rotor head)
FEL	Free Electron Laser
FELT	Free Electron Laser Technology
FEL-TNO	TNO Physics & Electronics Laboratory (Netherlands)
FEM	Force Effectiveness Measure
FEMA	Federal Emergency Management Agency (USA)
FES	Force Entry Switch. Communications
FET	Field Effect Transistor
FEW	Fighter Escort Wing
FEWS	Follow-on Early Warning System
FEWSG	Fleet Electronic Warfare Support Group
FF	Frigate (USA)
FFA	Aeronautical Research Institute of Sweden
FFA	Forces Francaises en Allemagne. French Forces in Germany
FFAR	Folding-Fin Aircraft Rocket
FFAR	Forward-Firing Aircraft Rocket
FFAR	Free-Flight Aircraft Rocket
FFB	Filter-Fan-Battery
FFC	Full Function Crew station
FFCS	Free-Fall Control System
FFG	Flensburger Fahrzeugbau GmbH (Germany)
FFG	Guided Missile Frigate (USA)
FFISTS	Fast Frigate Integrated Shipboard Tactical System (USA)
FFL	Light frigate or corvette
FFLAV	Future Family Light Armoured Vehicle programme (UK)
FFLG	Guided missile light frigate or corvette
FFOS	Forward Flying Observation System
FFPB	Free-Fall Practice Bomb
FFR	Fitted For Radio / Radar
FFR	Free Flight Rocket
FFR	Free Flooding Projector. Acoustic projector
FFSK	Fast Frequency Shift Keying
FFSP	Full Function Signal Processor
FFT	Fast Fourier Transform
FFTA	The Finnish Foreign Trade Association
FFVV	Federation Francaise de Vol a Voile (France)
FG	Field Gun
Fg Off	Flying Officer
FGA	Fighter Ground Attack
FGCP	Flight Guidance Control Panel
FGR	Fighter, Ground Attack Reconnaissance (UK)
FH	Field Howitzer
FH	Frequency Hopping
FHFS	Foundation Health Federal Services (USA)
FHI	Fuji Heavy Industries Ltd (Japan)
FHS	Forsvarshogskolan (Sweden)

FHU	Force Helicopter Unit (NATO)
FIBUA	Fighting In Built-Up Areas
FIFV	Future Infantry Fighting Vehicle (USA)
FIG	Fighter Interceptor Group
FII	Fuerzas de Intervencion Inmediata (Spain)
FILA	Fighting Intruders at Low Altitude
FIMS	Ferranti Integrated Mine Countermeasures System
FIPS	Federal Information Processing Standard
FIR	Flight Information Region
FIR	Forza d'Intervento Rapido. Rapid intervention force (Italy)
FIR	Functional Item Replacement
FIRAMS	Flight Incident Recorder & Aircraft Monitoring System
FIRMU	Flight Incident Recorder Memory Unit
FIS	Fighter Interceptor Squadron
FIS	Front Islamique de Salut (Algeria)
FIS3	Ferranti Integrated Submarine Sonar Suite
FISCS	Ferranti Integrated Submarine Combat Suite (UK)
FIST	Facility for Infantry Situation Training
FIST	Fire Support Team
FIST DMD	Fire Support Team Digital Message Device
FISTV	Fire Support Team Vehicle (USA)
FITOW	Further Improved TOW (USA)
FITS	Falkland Islands Trunk System
FIVEATAF	Fifth Allied Tactical Air Force (NATO)
FJ	Fast Jet
FJC	Falcon Jet Corp. (USA)
FKCU	Frequency & Key Copy Unit
FKMS	Frequency & Key Management System
FKMU	Frequency & Key Management Unit
FL	Flexible
FLA	Future Large Aircraft
FLAB	Flag Lieutenant to the Admiralty Board (UK)
FLAG	Flemish Aerospace Group (Belgium)
FLAGE	Flexible Lightweight Agile Guided Experiment
FLAIR	Field Low Altitude Intermediate-range Radar (France)
FLASH	Folding Light Acoustic System for Helicopters
FLAV	Family of Light Armoured Vehicles
FLCH	Flechette
FLET	Forward Line of Enemy Troops
FLIR	Forward Looking Infra-Red
FLN	Front de Liberation Nationale (Algeria)
FLOPSY	Fleet Operations Programming System
FLOT	Forward Line of Own Troops
FLPT	Fork Lift Pallet Trailer
FLS	Force Level Simulation
Flt	Flight
Flt Lt	Flight Lieutenant
FLTBDCST	Fleet Broadcast (US Navy)
FLTSATCOM	Fleet Satellite Communications (USA)
FM	Fabricaciones Militares (Argentina)

FM	Frequency Modulation
FM	Titanium tetrachloride
FMA	Fabrica Militar de Aviones SA (now Fabrica Argentina de Material Aerospacial) (Argentina)
FMA	Flight Mode Annunciator
FMBT	Future Main Battle Tank
FMC	Food Machinery Corporation (USA)
FMC	Fully Mission-Capable (USAF)
FMCS	Flight Management Computer System
FMCW	Frequency Modulated Continuous Wave
FMECA	Failure Mode, Effect & Criticality Analysis
FMF	Fleet Marine Force
FMGC	Filtre Mixte de Grande Capacite (France)
FMGC	Flight Management & Guidance Computer
FMICW	Frequency Modulated Interrupted Continuous Wave
FMJ	Full Metal Jacket
FMK	Flyvematerielkommandoen (Denmark)
FMLN	Farabundo Marti National Liberation Front (El Salvador)
FMO	Federalni Ministerstvo Obrany (Czechoslovakia)
FMOP	Frequency Modulation On Pulse
FMS	Ferranti Modular Sonar
FMS	Field-Maintenance Squadron (USAF)
FMS	Fire Marking System
FMS	Flight Management System
FMS	Foreign Military Sales
FMTV	Family of Medium Tactical Vehicles
FMU	Feeder Matching Unit. Harmonic filtering
FMU	Fleet Maintenance Unit
FMV	Forsvarets Materielverk (Sweden)
FMV	Swedish defence material administration
FN	Fabrique Nationale (Belgium)
FNF	Flash Non-Fragmentation
FNH	Fuerza Naval Hondurena (Honduras)
FNMI	FN Manufacturing Inc. (USA)
FNNH	Fabrique Nationale Nouvelle Herstal (Belgium)
FNSS	FMC-Nurol Savunma Sanayii (Turkey)
FO	Fibre Optics
FO	Flag Officer (Royal Navy)
FO	Follow-On
FO	Forward Observer
FOAC	Flag Officer, Aircraft carrier
FOAS	Fleet Operational Analysis Staff
FOB	Forward Operating Base
FOC	Full Operational Capability
FOCA	Fibre Optic Cable Assembly
FOCAS	Fibre Optical Communications for Aerospace Systems (USA)
FOCAS	Flag Officer, Carriers & Amphibious Ships (UK)
FOCOS	Foam Overhead Cover Support System
FOCSLE	Fleet Operational Command System Life Extension
FOD	Foreign Object Damage

FODL	Fibre-Optic Data Link
FODT	Fibre Optics Data Transmission
FODU	Fibre Optic Distribution Unit
FOF3	Flag Officer 3rd Flotilla (Royal Navy)
FOFA	Follow-On Forces Attack
FOG	Fibre Optics Guidance
FOG-M	Fibre Optic Guided Missile
FOHMD	Fibre-Optic Helmet-Mounted Display
FOI	Follow-On Interceptor (USAF)
FOL	Fly-Off Lever
FOL	Forward Operating Location
FOLD	Forward Observer Laser Designator
FOLPEN	Foliage Penetration
FOM	Forces d'Outre Mer (France)
FOMP	Fiber Optic Mortar Projectile (USA)
FONA	Flag Officer, Naval Aviation
FONAC	Flag Officer, Naval Air Command
FOO	Forward Observation Officer
FOP	Fleet Operations Programme
FOPS	Falling Object Protective Structure
FORACS	Fleet Operational Readiness Accuracy & Check System
FORATOM	Forum Atomique Europeen
FORTIS	Forward Observation & Reconnaissance Thermal Imaging System (Switzerland)
FORTRAN	Formula Translating System
FOSF	Flag Officer, Surface Flotillas
FOSIC	Fleet Ocean Surveillance Information Centre (US Navy)
FOSM	Flag Officer, Submarines
FOSNI	Flag Officer, Scotland & Northern Ireland (Royal Navy)
FOSSN	Follow-On Nuclear Submarine
FOST	Flag Officer, Sea Training (Royal Navy)
FOST	Force Oceanique Strategique (France)
FOTL	Follow-On To Lance
FOTS-LH	Fibre Optic Transmission System – Long Haul
FOURATAF	Fourth Allied Tactical Air Force (NATO)
FOV	Field-Of-View
FOW	Family Of Weapons
FP	Frangible Projectile
FPA	Focal-Plane Array
FPAS	Focal-Plane Array Seeker
FPB	Fast Patrol Boat
FPDA	Five Power Defence Agreement
FPF	Final Protective Firetask
FP-IMS	Fixed Point Ion Mobility Spectrometer chemical detector
FPIWA	First-Pass In-Weather Attack
FPLM	Mozambique Army
FPMU	Fuel Pump Monitoring Unit
FPP	Fleet Planning & Programming
FPR	Federal Procurement Regulation (USA)
FPR	Fibre Reinforced Plastic

FPS	Fast Packet Switch. Network communications
FQIS	Fuel Quantity Indication System
FRAG	Fragmentation
FRAM	Fleet Rehabilitation And Modernisation programme (US Navy)
FRAS	Free-Rocket Anti-Submarine
FRC	Fibre Reinforced Composite
FRC	Flare Remote Control
FRG	Federal Republic of Germany
FROG	Free-Rocket-Over-Ground
FRP	Fibre Reinforced Plastic
FRPA	Fixed Reception Pattern Antenna
FRS	Fighter, Reconnaissance, Strike (UK)
FRS	Fleet Readiness Squadron
FRUD	Front for Restoration of Unity & Democracy (Djibouti)
FRUSA	Flexible Rolled-Up Solar Array
FRV	Final Rendezvous
FS	Feasibility Study
FS	Field Standard
FS	Fighter Squadron
FS	Fin-Stabilised
FS	Firing Section for missile system
FS3	Future Strategic Strategy Study
FSA	Field Support office (USA)
FSA	Future Surface-to-Air
FSAA	Flight Simulator for Advanced Aircraft
FSAF	Future Surface-to-Air Family of weapon systems
FSAGA	First Sortie After Ground Alert
FSAS	Fuel Savings Advisory Systems
FSAT	Full-Scale Aerial Target
FSA/CAS	Fuel Savings Advisory & Cockpit Avionics System
FSB	Faltschwimmbrucke (Germany)
FSB	Forward Support Battalion
FSC	Fuel Savings Computer
FSCV	Fire Support Combat Vehicle
FSD	Full-Scale Deflection
FSD	Full-Scale Development
FSDS	Fin-Stabilised Discarding Sabot
FSE	Fire Support Element
FSED	Full-Scale Engineering Development (USA)
FSG	Field Support Group
FSI	Fastening Systems International Inc. (USA)
FSIC	Forward Sensor Interface & Control module
FSK	Frequency Shift Keying
FSMTC	Full-Size Moving Tank Target Carrier
FSO	Fire Support Officer (USA)
FSP	Fragment Simulator Projectile
FSP/UPT	Flight-Screening Program & Under-graduate Pilot Training (USAF)
FSR	Flood Search Routeing. Communications
FSR	Flying Selection Squadron (Royal Air Force)
FSRS	Frequency Selective Receiver System

FSS	Field Support Subsystem (UAV)
FSS	Fire Sensing & Suppression system
FSS	Fire Support Section
FST	Future Soviet Tank / Follow-on Soviet Tank (CIS)
FSU	Fleet Support Unit
FSUP	Flight Simulator Update Program
FSV	Fire Support Vehicle
FSW	Forward-Swept Wing
FT	Follow Through
FT	Force Terminal. Multiple-channel communications terminal (USA)
FTA	Frangible Training Ammunition
FTAAS	Fast Time Acoustic Analysis System (Australia)
FTCA	Future Tactical Combat Aircraft
FTD	Flight Training Device
FTD	Foreign Technology Division (USAF)
FTG	False Target Generator
FTI	Fixed Target Imagery
FTLO	False Target Lock-On
FTMA	Future Tank Main Armament
FTP	Fly-To Point
FTP	Fusion Track Processor
Ftr	Fighter
FTR	False Target Rejection
FTS	Factory-based Test Stand
FTS	Flexible Turret System
FTS	Future Tank Study
FTSS	Ferranti-Thomson Sonar Systems UK Ltd
FTV	Functional Test Vehicle
FTX	Field Training Exercise (UK)
FU	Firing Unit
Fu H	Fuhrungsstab des Heeres (Germany)
Fu L	Fuhrungsstab der Luftwaffe (Germany)
Fu M	Fuhrungsstab der Marine (Germany)
FUDS	Formerly Used Defence Sites
FUG	Felderito Uso Gepkosci. Amphibious scout car (Hungary)
FUP	Form-Up Point (UK)
FUSEP	Fuerzas de Seguridad Publica (Honduras)
FUV	Far Ultra-Violet
FV	Field Vehicle
FV	Fighting Vehicle
FVRDE	Fighting Vehicles Research & Development Establishment
FVS	Fighting Vehicle System (USA)
FVSC	Fighting Vehicle Systems Carrier (USA)
FV/GCE	Fighting Vehicle / Gun Control Equipment
FWA	Fleet Weapon Acceptance
FWAM	Full Width Attack Mine
FWD	Four-Wheel Drive
FWE	Foreign Weapons Evaluations
FWETE	Foreign Weapons Equipment Technology Evaluation (USA)
FWM	Feinmechanische Werke Mainz GmbH (Germany)

FWOC	Fleet Weather & Oceanographic Centre / Computer
FWOCCER	Fleet Weather & Oceanographic Centre Computer Equipment Replacement
FWS	Fighter Weapons School (USAF)
FWT	Fleetwork Trainer
FXA	Fleet Exercise Area
FY	Fiscal Year
FYDP	Five-Year Defence Programme
F&C	Fire & Control (missile guidance)
F&F	Fire & Forget (missile guidance)
F&U	Fire & Update (missile guidance)

G

G	Gendarmerie (France)
G	Nerve agent. Chemical warfare
GA	Gun Assembly
GA	Tabun. Chemical warfare nerve agent
GaAs	Gallium Arsenide
GACC	Ground Attack Control Capability
GAIC	Guizhou Aviation Industry Corp. (China)
Gambit	General Anti-Material Bomblet with Improved Terminal effects
GAO	General Accounting Office (USA)
GAP	Glycidyl Azide Polymer
GAP	Gun Aiming Post
GAPA	Ground-to-Air Pilotless Aircraft
GAR	Generic Airborne Radar
GAR	German Air Force
GAR	Grand Army of the Republic (USA)
GARDIAN	General Area Defense Integrated Anti-missile laser weapon (USA)
GARIM	Air Group Headquarters (Mauritania)
GAR/I	Ground Acquisition Receiver / Interrogator
GAS	General Air Situation (NATO)
GASC	Ground Air Support Command (USA)
GASS	General Air & Surface Situation (NATO)
GATT	General Agreement on Tariffs & Trade
GAU	Gun Aircraft Unit
GB	Sarin. Chemical warfare nerve agent
GBCS-H	Ground-Based Common Sensor for Heavy divisions
GBCS-L	Ground-Based Common Sensor, Light
GBFEL	Ground-Based Free Electron Laser
GBHE	Ground-Based Hypervelocity Experiment
GBHRG	Ground-Based Hypervelocity Rail Gun
GBI	Ground-Based Interceptor
GBL	Ground-Based Laser
GBMD	Global Ballistic Missile Defence
GBR	Ground-Based Radar
GBRT	Ground-Based Radar Terminal
GBS	Ground-Based Sensor
GBU	Glide Bomb Unit
GCA	Ground Controlled Approach (radar)
GCB	Gun Control Box
GCC	Co-operation Council for the Arab States of the Gulf
GCDU	Gunner Control & Display Unit
GCE	General Consulting & Engineering Srl. (Italy)
GCE	Gordon Consulting Engineers Ltd (Israel)
GCE	Gun Control Equipment
GCHQ	Government Communications Headquarters
GCI	Ground-Controlled Interception
GCM	General Court-Martial
GCR	Ground Clutter Reduction

GCS	Ground Control Station
GCS	Guidance & Control Section
GCT	Government Competitive Test
GCT	Grande Cadence de Tir. Self-propelled gun (France)
GCU	Ground Control Unit (UAV)
GCW	Gross Combination Weight
GD	General Duties (UK)
GD	General Dynamics Corp. (USA)
GD	Graphic Display
GD	Soman. Chemical warfare nerve agent
GDAP	Government Document Application Profile
GDE	Graphics Differential Engine
GDI	Graphic Datakits International (USA)
GDIP	General Defence Intelligence Programme office (USA)
GDL	Gas Dynamic Laser
GDMS	Geographic Data Management System
GDOP	Geometric Dilution Of Precision
GDP	General Defence Plan
GDP	Gross Domestic Product
GDRC	Ground Defence Reporting Cell
GDS	Gun Display System
GDSC	General Dynamics Services Co. (USA)
GDT	Ground Data Terminal
GDU	Graphics Display Unit
GDU	Gun Display Unit
GE	General Electric Co. (USA)
GE AMERICOM	GE American Communications Inc.
GEADGE	German Air Defence Ground Environment
GEAMOS	Gefechtsaufklarungsmittel und Ortungssytem (Germany)
GEC	The General Electric Co. plc (UK)
GEDU	Gun Elevation Displacement Unit
GEE	Ground Exploration System
GEHOC	German HAWK Operations Centre
GEM	GPS Embedded Module
GEM	Graff Electronic Machines Ltd (UK)
GEM	Graphite Epoxy Motor
GEMAG	General Electric Mobile Air Defense Gun
GEMSS	Ground-Emplaced Mine-Scattering System
Gen	General
GEN II	Second Generation
GENSER	General Service
GENTEL	General Intelligence (NATO)
GEN-X	Generic Expendable
GEODSS	Ground-based Electro-Optical Deep Space Surveillance
GEOREF	World Geographic Reference system (USAF)
GES	Ground Earth Station
GETEN	Research & Development Centre (Hellenic Navy)
GETS	Generic ESM Test Set
GETSCO	General Electric Technical Services Co. Inc. (Germany)
GEV	Ground Effect Vehicle

GFAE Government-Furnished Avionic Equipment
GFCC Gunner's Fire Control Console
GFCS Gun Fire Control System
GFE Government Furnished Equipment
GfK Glassfibre-reinforced plastic (Germany)
GFRP Glassfibre-Reinforced Plastic
GGP GPS Guidance Package
GGS Gyro Gunsight
GGT Gerber Garment Technology (USA)
GH Gun / Howitzer
GHQ General Headquarters
GHT Gesellschaft fur Hochdrucktechnik MBH (Germany)
GI Government Issue. (US Army)
GI Groupement de l'Instruction (Switzerland)
GI Gunnery Instructor (Royal Navy)
GIA Garber International Associates Inc. (USA)
GIAT Groupement Industriel des Armements Terrestres (France)
GIBMED Gibraltar Mediterranean Command (NATO)
GIC Glatzer Industries Corp. (USA)
GIE Groupement d'Interet Economique (France)
GIFAS Groupement des Industries Francaises Aeronautiques et Spatiales
GIM Ground Influence Mine
GIS Geographic Information System
GISS Goddard Institute for Space Studies (USA)
GIST General Intelligence Ship Terminal
GITS GM Hughes Electronics Integrated TOW Sight
GIUK Greenland / Iceland / United Kingdom (NATO)
GKS Graphic Kernel System
GLAADS Gun Low Altitude Air Defense System (USA)
GLC Gun Lay Computer
GLCM Ground-Launched Cruise Missile
GLH-H Ground-Launched Hellfire-Heavy (USA)
GLLD Ground Laser Locator Designator (USA)
GLO Ground Liaison Officer (UK)
GLOC G-induced Loss Of Consciousness
GLS Gesellschaft fur Logistischen Service (Germany)
GM Guided Missile
GMC General Motors Corp. (USA)
GMC Ground Movement Controller
GMD Guided Missile Destroyers (CIS)
GMDL GEC-Marconi Dynamics Ltd (UK)
GMF Ground Mobile Forces
GMFSC Ground Mobile Forces Satellite Communications
GMGN Guided Missile Group Netherlands
GMHE GM Hughes Electronics Corp. (USA)
GMLS Guided Missile Launch System
GMOTS Gun Maintenance & Operational Training System
GMR Ground Mapping Radar
GMRT Ground-Mapping Radar Training
GMS GEC-Marconi Systems Pty Ltd (Australia)

GMS	Gun Management System
GMT	Greenwich Mean Time
GMT	Greenwich Meridian Time
GMTI	Ground Moving-Target Indicator
GMV	Grupo de Mecanica del Vuelo SA (Spain)
GMVLS	Guided Missile Vertical Launching System
GMWS	Guided Missile Weapon System
GMZ	Gusenichniy Minniy Zagraditel. Tracked Minelayer (CIS)
GNATS	Graphic Navlink Aircraft Tracking System
GNC	General Navigation Computer
GNC	Guidance / Navigation / Control
GNCS	Guidance Navigation Control System
GNP	Gross National Product
Gnr	Gunner
GNS	Global Navigation System
GNS	Gyrocompass Navigation System
GO	Gunnery Officer (Royal Navy)
GOC	General Officer Commanding
GOCO	Government-Owned, Contractor-Operated (USA)
GOE	Grupo de Operaciones Especiales. Special forces group (El Salvador)
GOLD	General On-Line Diagnostic
GONS	Guns Orientation & Navigation System
GOP	General Operational Plot
GORU	Gun Order Responder Unit
GOS	General Operator Station
GP	General-Purpose
GP	Guided Projectile
Gp Capt	Group Captain
GPALS	Global Protection Against Limited Strikes
GPC	Greengate Polymer Coatings Ltd (UK)
GPC	Secretariat of the General People's Congress (Libya)
GPCDU	General Purpose Control & Display Unit
GPDC	General Purpose Digital Computer
GPEOD	General Purpose Electro-Optical Director
GPFU	Gas-Particulate Filter Unit
GPI	Glide Path Indicator
GPIB	General Purpose Interface Bus (or Board)
GPM	Gepanserte Pioniermaschine (Germany)
GPMG	General-Purpose Machine Gun
GPO	Gun Position Officer
GPP	Groupement pour les Gros Propulseurs a Poudre (France)
GPS	Global Position System (formerly NAVSTAR)
GPS	Gunner's Primary Sight
GPSSU	Global Positioning System Sensor Unit
GPTB	Gunpell Target Board
GPTTS	Gunner's Primary Tank Thermal Sight
GPU	Ground Power Unit
GPU	Gun Pod Unit
GPWS	Ground Proximity Warning System
gr	Gunner

GR	General Regulator (USA)
GR	Ground attack Reconnaissance
GRC	General Research Corp. (USA)
GRD	Gruppe fur Rustungdienste (Germany)
GRDS	Generic Radar Display System
GRE	Ground Read-Out Equipment
GRP	Glass-reinforced plastic
GRSC	Ground Radio Servicing Centre
GRSF	Ground Radio Servicing Flight (Royal Air Force)
grt	Gross registered tonnage (ship)
GRU	Gurkha Reserve Unit (Brunei)
GS	General Support
GS	Glide Slope
GS	Ground Speed
GSA	Gunsight, Surface-to-Air
GSC	General Service Cargo Land Rover variant (UK)
GSE	Global Security Environment
GSE	Ground Support Equipment
GSFC	Goddard Space Flight Center (USA)
GSFG	Group of Soviet Forces in Germany (NATO)
GS-FlGr	Grenzschutz-Fliegergruppe (Germany)
GSM	Granite State Manufacturing (USA)
GSM	Ground Station Module
GSP	Gusenichniy Samokhodniy Parom. Heavy Amphibious Ferry (CIS)
GSR	General Staff Requirement (USA)
GSR	Ground Surveillance Radar
GSRS	General Support Rocket System
GSS	Generalised Simulation / Stimulation system
GSS	Ground Support System
GSSA	General Support Supply Activity (USA)
GSSS	Gyro-Stabilised Sighting System (Israel)
GST	General Staff Target
GST	Gesellschaft fur System-Technik (Germany)
GSTS	Ground-Based Surveillance & Tracking System
GT	Geared Turbine
GT	Gun Tank
GTA	Ground Test Accelerator
GTACS	Ground Target Attack Control System
GTC	Greene, Tweed & Co. Ltd (UK)
GTCS	Gun Test & Control System
GTD	Gun Turret Drive
GTECC	GTE Communications Corp. (USA)
GTEISC	GTE International Systems Co. (USA)
GTFM	Generalised Tamed Frequency Modulation
GTI	German Tank Improvement
GTNC	German Territorial Northern Command
GTPE	Gun Time Per Engagement
GTRE	Gas Turbine Research Establishment (UK)
GTS	Gas Turbine Starter
GTTC	Ground Tactical Training Centre

GTV	Glide Test Vehicle
GTV	Ground / Guidance Test Vehicle
GUI	Graphical User Interface
GV	Giant Viper
GVSC	Generic VHSIC Spaceborne Computer (USAF)
GVW	Gross Vehicle Weight
GW	Guided Weapon
GWEN	Ground Wave Emergency Network (USA)
GWS	Gatling Weapon System
GWS	Global Wulfsberg Systems (USA)
GWS	Guided Weapon System
GWST	Gabriel Weapon System Trainer
GZ	Ground Zero
G&C	Guidance & Control
G/M	Gun / Missile selector
g/t	Gain / temperature
G/VLLD	Ground / Vehicular Laser Locator Designator. Pronounced 'GLID'

H

H	Hydrographer to the navy (UK)
H	Mustard. Chemical warfare blister agent
HA	Heavy Artillery
HAARS	High-Altitude Airdrop Resupply System
HAAS	High Altitude Active / semi-active Seeker
HAB	Heavy Assault Bridge (USA)
HAC	Honourable Artillery Company (UK)
HAC	House Appropriations Committee (USA)
HAC	Hover / Approach Coupler
HAC	Hughes Aircraft Co. (USA)
HACJ	Helicopter Applique Communications & Jammer
HACV	Heavy Armament Combat Vehicle
HADAS	Helmet Airborne Display & Sight
HADR	Hughes Air Defence Radar
HADS	Helicopter Air Data System
HAECO	Hong Kong Aircraft Engineering Co. Ltd
HAFCS	Howitzer Advanced Fire Control System
HAHST	High Altitude High-Speed Target
HAI	Hellenic Aerospace Industry Ltd (Greece)
HAINS	High Accuracy Inertial Navigation System
HAIR	High Altitude IR
HAIS	Hydro Acoustic Information System
HAISCO	Hughes Aircraft International Service Co. (USA)
HAISS	High Altitude IR Sensor System
HAL	Hindustan Aeronautics Ltd (India)
HALE	High Altitude / Long Endurance (UAV)
HALO	High Altitude Low Observable
HALO	High Altitude, Low Opening (parachuting)
HAMC	Harbin Aircraft Manufacturing Co. (China)
HAP	Helicoptere d'Appui Protection (France)
HAPS	Helicopter Acoustic Processing System
HAR	Helicopter, search & rescue (UK)
HARD	Helicopter & Airplane Radio / Radar Detection
HARM	High-Speed Anti-Radiation Missile
HARP	High Altitude Research Programme
HARS	Heading / Attitude Reference System
HARU	Heading & Attitude Reference Unit
HAS	Hardened Aircraft Shelter
HAS	Helicopter, anti-submarine (UK)
HAS	Hood, Aircrew Survival
HAS	Hover-Augmentation System
HASC	House Armed Services Committee (USA)
HAST	Harrier Avionics System Trainer
HAST	High Altitude Supersonic Target
HAT	Harbour Acceptance Testing
HATIS	Helmet Acquisition & Tracking Indication System
HATS	HAWK Advanced Training Simulator

HATS	Helicopter Automatic Targeting System
HAW	Heavy Anti-tank Weapon
HAWC	Homing And Warning Computer
HAWFCAR	Helicopter Adverse Weather Fire Control Radar
HAWK	Homing-All-the-Way-Killer (missile)
HAWK-PIP	HAWK Product Improvement Program
HAWK/HIP	HAWK Improvement Program
HAWTADS	Helicopter All-Weather Target Acquisition & Destruction System
HB	Heavy Barrel
HB	Hollow Base. Ammunition
HBAR	Heavy Barrelled Automatic Rifle
HC	Hard Core. Ammunition
HC	Helicopter, cargo (UK)
HC	Hexachloroethane-zinc smoke mixture
HC	High Capacity
HCE	Hexachloroethane
HCER	High Capacity Extended Range
HCF	Honorary Chaplain to the Forces
HCHE	High Capacity High Explosive
HCI	Hughes Communications Inc. (USA)
HCI	Human Computer Interface
HCN	Hydrogen Cyanide. Chemical warfare blood agent (also known as AC)
HCPE	Hybrid Collective Protection Equipment
HCT	Heavy Crawler Tractor
HCT	Helicopter Control Trainer
HCT	HOT Compact Turret
HCU	Hoist Control Unit
HD	Distilled Mustard. Chemical warfare blister agent
HD	High Drag (bomb)
HDB	High Density Bombing
HDBK	Hand Book
HDD	Head-Down Display
HDDR	Head-Down Display Radar
HDFPA	High Density Focal Plane Array
HDI	High-Density Interconnect technique
HDLC	High Level Data Link Control
HDOS	Hughes Danbury Optical Systems Inc. (USA)
HDRV	Heavy Duty Recovery Vehicle
HDS	Headquarters Distribution System
HDTM	Heavy-Duty Tank Target Mechanism
HDU	Hose Drum Unit
HDW	Howaldtswerke Deutsche Werft AG (Germany)
HE	High Explosive
HEA	Hydrogen Engineering Applications Ltd (UK)
HEAA	High Explosive, Anti-Armour
HEAD	High Explosive, Air Defence
HEAP	High Explosive, Anti-Personnel
HE-APERS-FRAG	High Explosive, Anti-Personnel, Fragmentation
HEAP-T	High Explosive, Anti-Personnel, Tracer
HEAT	High Explosive, Anti-Tank

HEAT	Hostile Expendable Aerial Target
HEAT-FS	High Explosive, Anti-Tank, Fin-Stabilised
HEAT-MP	High Explosive, Anti-Tank, Multi-Purpose
HEAT-MP(P)	High Explosive, Anti-Tank, Multi-Purpose, (Practice)
HEAT-T	High Explosive, Anti-Tank, Tracer
HEAT-T-HVY	High Explosive, Anti-Tank, Tracer, Heavy
HEAT-T-MP	High Explosive, Anti-Tank, Tracer, Multi-Purpose
HEAT-TP-T	High Explosive, Anti-Tank, Target, Practice, Tracer
HECO	Heckethorn Manufacturing Co. (USA)
HEDI	High Endoatmospheric Defence Interceptor
HEDP	High Explosive, Dual Purpose
HEDS	High Endoatmospheric Defense System
HEER	High Explosive, Extended Range (USA)
HE-FRAG	High Explosive, Fragmentation
HE-FRAG-FS	High Explosive, Fragmentation, Fin-Stabilised
HE-FS	High Explosive, Fin-Stabilised
HEFT	High Explosive, Follow Through
HEI	High Explosive, Incendiary
HEI-SAP	High Explosive, Incendiary, Semi-Armour-Piercing
HEIT	High Explosive, Incendiary, Tracer
HEL	High-Energy Laser
HEL	Human Engineering Laboratory (USA)
HELBAT	Human Engineering Laboratory Battalion Artillery Test
HELCM	High Energy Laser Countermeasures
HELD	Helicopter Laser Designator
HELEX	High Energy Laser Experimental
HELIBRAS	Helicopteros do Brasil S/A
HELIOS	Helicopter Instrument & Operational Procedures Simulator
HELMS	Helicopter Malfunction System
HELP	Howitzer Extended Life Program (USA)
HELPS	Helicopter Protection & Support
HELRAPS	Heliborne Long-Range Acoustic Path Sonar
HELRAS	Helicopter Long-Range Sonar
HELSTF	High Energy Systems Test Facility
HELTADS	High Energy Laser Tactical Air Defence System
HEMA	Heavy Engineering Manufacturers' Association (Australia)
HEMAT	Heavy Expanded Mobility Ammunition Trailer (USA)
HEMC	High Explosive, Medium Capacity
HEMLOC	Helicopter Emitter / Locator Countermeasures
HEMP	High Altitude EMP protection
HEMTT	Heavy Expanded Mobility Tactical Truck (USA)
HEOB	Hostile Electronic Order of Battle
HE-OM	High Explosive OTO Munition
HEP	High Explosive, Plastic
HEPD	High Explosive, Point Detonating
HEPI	High Explosive, armour Piercing & Incendiary
HEPL	High Energy Pulse Laser
HEP-T	High Explosive, Plastic, Tracer
HER	Harsh Environment Recorder
HERA	High Explosive, Rocket-Assisted

HERALD	Helicopter Equipment for Radar And Laser Detection
HERID	High Energy Railgun Integration Demonstration (USAF)
HERO	Hazards of Electro-magnetic Radiation to Ordnance
HEROS	Heeres-Fuhrungsinformationsystem zur Rechnergestutzten Operationsfuhrung in Staben (Germany)
HERT	Headquarters Emergency Relocation Team
HE-S	High Explosive, Spotting
HES	Hardcopy Exploitation Segment (JSIPS)
HE-SD	High Explosive, Self-Destroying
HESH	High Explosive, Squash Head
HESH-T	High Explosive, Squash Head, Tracer
HE-T	High Explosive, Tracer
HET	Heavy Equipment Transporter
HE-T SD	High Explosive, Tracer, Self-Destruct
HETA	Harpoon Engagement Training Aid
HET-PF	High Explosive, Tracer, Percussion Fuze
HEU	HUD Electronics Unit
HEU	Hull Electronics Unit
Hexal	Hexogen / aluminium powder
HE/AP	High Explosive, Armour Piercing
HE/DPSD	High Explosive, Dual Purpose, Self-Destruction
HE/PDSD	High Explosive, Point Detonating, Self-Destruction
HE/PR	High Explosive, Practice
HF	Harassing Fire
HFAC	Helicopter Flight Advisory Computer
HFAJ	High Frequency Anti-Jam
HFAJ	High Frequency Anti-Jam programme (USA)
HFCC	Howitzer Fire Control Computer (USA)
HFCS	Helicopter Fire Control System
HFHTB	Human Factors Howitzer Test-Bed (USA)
HFIC	HF Intra-task force Communications (USA)
HFIP	High Frequency Improvement Programme (NATO)
HFL	Hydrogen Fluoride Laser
HFM	Heavy Force Modernisation (USA)
HFMS	High Frequency Monitoring System
HFP	High Fragmentation Projectile
HFR	Height-Finder Radar
HGAS	High Gain Antenna System
HgCdTe	Mercury-cadmium-telluride
HGR	Hypervelocity Guided Rocket
HGS	Holographic Guidance System
HGV	Heavy Goods Vehicle
HHC	Hand-Held Controller
HHCU	Hand-Held Control Unit
HHLL	Hand-Held Laser Locator
HHLR	Hand-Held Laser Rangefinder
H-Hour	Time of commencement
HHTI	Hand-Held Thermal Imager
HHTR	Hand-Held Tactical Radar
HHUD	Holographic Head-Up Display

HIBAT	Helicopter Identification By Acoustic Techniques
HIBIRD	Helicopter Identification By Infra-Red Detection
HIBRAD	Helicopter Identification By Radar Detection
HIDEC	Highly Integrated Digital Engine Control
HIDS	Headquarters Information Distribution System
HIFR	Hover In-Flight Refuelling
HIGE	Hover In Ground Effect
HIK	Hook Interface Kit for transport of ISO containers
HILEX	High Level Exercise
HIM	High Inclination Mission
HIMAD	High to Medium Altitude Air Defence
HIMAG	High-Mobility Agility Test Vehicle (USA)
HIMAT	Highly Manoeuvrable Aircraft Technology programme
HIP	Howitzer Improvement Program (USA)
HIPACT	High Power Acoustic Coaxial Transducer
HIPAD	High Performance Air Defence
HIPAR	High Power Acquisition Radar
HIPAS	High Performance Sonar
HIPCOR	High-Power Coherent Radar
HIRAM	Hycor Infra-Red Anti-Missile decoy (USA)
HIRE	Hughes Infra-Red Equipment thermal subsystem
HIRIS	High resolution Infra-red System
HIRNS	Helicopter Infra-Red Navigation System
HIRS	Helicopter Infra-Red System
HIS	High Integrity Systems Ltd (UK)
HIS	Hostile Intelligence Service
HISOS	Helicopter Integrated Sonar System
HI-SPOT	High Altitude Surveillance Platform for Over-the-horizon Targeting
HISS	Horizon IR Surveillance Sensor (USA)
HIT	Homing Intercept Technology
HIT	Hughes Improved Terminal
HITMORE	Helicopter Installed Television Monitor Recorder
HITP	High Ignition Temperature Propellant
HITS	High-definition Indoor Trainer for Small-arms
HITS	Hostile Identification / Targetting System
HITS	Hughes Integrated Tank Sight (USA)
HITTS	Harpoon Interactive Tactical Training System
HIU	Heading Indicator Unit
HK	Heckler & Koch GmbH (Germany)
HKV	Hughes Kongsberg Vaapenfabrik (Norway)
HL	Mustard-Lewisite mixture. Chemical warfare blister agent
HLD	Head Level Display
HLG	High Level Group (NATO)
HLH	Heavy Lift Helicopter
HLLV	Heavy Lift Launch Vehicle
HLTF	High Level Task Force (NATO)
HLTF (R)	High Level Task Force (Reinforced) (NATO)
HLV	Heavy Lift Vehicle
HLVW	Heavy Logistic Vehicle Wheeled (Canada)
HLWE	Helicopter Laser Warning Equipment

HMAC	Hugo Marom Aviation Consultants Ltd (Israel)
HMACI	Hugo Marom Aviation Consultants International Inc. (USA)
HMAFV	Her Majesty's Air Force Vessel (UK)
HMAS	His / Her Majesty's Australian Ship
HMC	Howitzer Motor Carriage
HMCS	His / Her Majesty's Canadian Ship
HMD	Helicopter Mine Dispenser
HMD	Helmet Mounted Display
HMEP	Helmet-Mounted Equipment Platform
HMG	Heavy Machine Gun
HMH	Helmet-Mounted Head-Up-Display
HML	Hard Mobile Launcher
HMLC	High Mobility Load Carrier
HMMHE	High Mobility Material Handling Equipment
HMMWV	High Mobility Multi-purpose Wheeled Vehicle (USA)
HMMWV	High-Mobility Multi-purpose Wheeled Vehicle (Hummer) (USA)
HMP	Heavy Machine Gun Pod
HMRT	Heavy Material Recovery Team
HMRT	Heavy Mobile Repair Team vehicle (Canada)
HMS	Helmet-Mounted Sight
HMS	His / Her Majesty's Ship (Royal Navy)
HMS	Hull-Mounted Set
HMSE	HAWK Mobility, Survivability & Enhancement
HMTR	Highly Mobile Tactical Radar
HMTT	High Mobility Tactical Truck
HMU	Helmet-Mounted Unit
HMU	Hydro-Mechanical Unit
HMX	Cyclotetramethylenetetranitramine. Shock sensative explosive
HMX	Phlegmatised octol
HN	Nitrogen Mustard. Chemical warfare blister agent
HN3	Nitrogen mustard gas. Chemical warfare agent (USA)
HNS	Hughes Network Systems Inc. (USA)
HNVS	Helicopter Night Vision System
HOB	Height Of Burst
HOD	Head Of Department
HOD	Head-Out Display
HOE	Holographic Optical Element technology
HOE	Homing Overlay Experiment
HOFIN	Hostile Fire Indicator
HOGE	Hover Out of Ground Effect
HOJ	Home-On-Jam
HOMS	Homing Optical Missile System
HOPS	Helmet Optical Position Sensor
HORIZON	Helicoptere d'Observation Radar et d'Investigation sur Zone (France)
HOT	Harpoon Onboard Trainer
HOT	Haut subsonique Optiquement Teleguide tire d'un tube. High subsonic optically guided fired from a tube anti-tank missile (France)
HOT	High Operating Temperature (France)
HOT	High-subsonic Optically Teleguided
HOTAS	Hands On Throttle And Stick

HOTASA	Hands On Throttle And Stick Aid
HOTOL	Horizontal Take-Off & Landing
How	Howitzer
HOWD	Helicopter Obstacle Warning Device
HOWLS	Hostile Weapons Location System (USA)
HP	Hewlett-Packard Co. (USA)
HP	High Pressure
HP	Hollow Point
HPA	High Power Amplifier
HPA-1	High Performance Absorber – 1 (Finland)
HPAG	High Power Amplifier Group
HPC	High Pressure Compressor
HPD	Haut Pouvoir de Destruction (France)
HPD	Hertford Production Design Ltd (UK)
HPF	High Performance, Fragmentation
HPFP	High Performance Fragmentation Projectile
HPG	High Pressure Gun
HPI	High Power Illuminator
HPI	High Probability of Intercept
HPL	High Power Laser
HPLB	High Power Laser Blinding
HPM	High Power Microwave
HPO	Hanseatische Prazisions-und Orbittechnik GmbH (Germany)
HPS	Helmet Pointing System (or Sight)
HPT	Height-Pressure Test
HPT	High Power Transmission set
HPT	High Pressure Turbine
HPZ	Helicopter Protected Zone
HQ	Headquarters
HQ SACLANT	Headquarters of the Supreme Allied Commander, Atlantic (NATO)
HQMSS	Headquarters Mission Support System
HQSTC	Headquarters RAF Strike Command
HQU	Headquarters Unit
HQUNTAC	Armed Forces Headquarters (Cambodia)
HRF	Heavy Rebuild Factory
HRG	Hemispherical Resonator Gyro
HRIR	High Resolution Ifra-red Radiometer
HRLR	High Repetition Laser Rangefinder
HRM	Herederos de Ramon Mugica SA (Spain)
HRM	Hrvatska Ratna Mornarica. Croatian Navy
HRN	High Resolution Navigation mode
HRP	High Risk Personnel
HRR	High Resolution Radar
HS	Hardware Specialty Co. Inc. (USA)
HS	Hartman Systems (USA)
HS	Helikopter Service A/S (Norway)
HSDB	High Speed Data Bus
HSDE	Hawker Siddeley Dynamics Engineering Ltd (UK)
HSG	Hawker Siddeley Group plc (UK)
HSI	Horizontal Situation Indicator

HSIC	High Speed Integrated Circuit
HSM	Hard Structure Munition
HSOS	Helicopter Stabilized Optronic Sight (South Africa)
HSPP	Hawker Siddeley Power Plant Ltd (UK)
HSRS	High Speed Resetting Switches
HSS	Helicopter Support Ship
HSSL	Helicopter Self-Screening Launcher
HSSS	Helicopter Secure Speech System
HSTV(L)	High Survivability Test Vehicle (Lightweight) (USA)
HT	Mustard-T mixture. Chemical warfare blister agent
HTADS	Helmet Target Acquisition & Designation System
HTC	Hard Target Capability
HTI	Horizons Technology Inc. (USA)
HTI	Hughes Training Inc. (USA)
HTKP	Hard-Target Kill Probability
HTNS	Helicopter Tactical Navigation System
HTOVL	Horizontal Take-Off, Vertical Landing
HTP	High Test Peroxide
HTPB	Hydroxy-Terminated-Poly-Butadiene (propellant)
HTSC	High-Temperature Superconductor (USA)
HTT	Heavy Tracked Tractor
HTTB	High Technology Test Bed (USA)
HTU	Hand-held Terminal Unit
HUD	Head-Up Display
HUDC	Head-Up Display Computer
HUDWAC	Head-Up Display & Weapon Aiming Computer
HUDWASS	Head-Up Display Weapon-Aiming Subsystem
HUMC	Health & Usage Monitoring Computer
HUMINT	Human Intelligence
HUMS	Health & Usage Monitoring System
HUTTS	Hayes Universal Tow Target System
HV	High Velocity
HVA	High Value Unit
HVAP	High Velocity, Armour-Piercing
HVAPDS-T	High Velocity, Armour-Piercing, Discarding Sabot, Tracer
HVAPFSDS	High Velocity, Armour-Piercing, Fin-Stabilised, Discarding Sabot
HVAP-T	High Velocity, Armour-Piercing, Tracer
HVAR	High Velocity Aircraft Rocket
HVBP	High Velocity Ballistic Protection
HVD	Helmet Visor Display
HVG	Hypervelocity Gun
HVIT	High Volume Information Transfer (USA)
HVL	Hypervelocity Launcher
HVM	High Velocity Missile
HVM	Hypervelocity Missile
HVMS	High Velocity Medium Support weapon system
HVMS	Hypervelocity Support Weapon
HVSD	High Value Site Defense (USA)
HVSS	Horizontal Volute Spring Suspension
HVTA	High Value Target Acquisition

HVTP-T	High Velocity, Target Practice, Tracer
HVY	Heavy
HYWAYS	Hybrids With Advanced Yield for Surveillance programme, to produce LWIR focal plane arrays (USA)
H/TEU	Hull & Turret Electronics Unit

I

I2E	Etudes Electroniques Informatiques (France)
I2F	Intelligent Influence Fuze
I2R	Imaging Infra-Red
I2S	Infra-red Imaging System
IA	Immediate Action
IA	Irrespirable Atmosphere
IAAFA	Inter-American Air Forces Academy (USA)
IABG	Industrieanlagen-Betriebsgesellschaft mbH (Germany)
IAC	Industrial Acoustics Company Ltd (UK)
IAC	Integrated Avionics Computer
IAC	Intelligence Analysis Center (USA)
IACS	Integrated Avionics Control System
IADB	Inter-American Defense Board, US delegation (USA)
IADC	Inter-American Defense College
IADGES	Indian Air Defence Ground Environment System
IADS	Iceland Air Defence System
IADS	Integrated Air Defence System
IADT	Integrated Automatic Detection & Tracking
IAE	Institute for the Advancement of Engineering (USA)
IAEA	International Atomic Energy Agency
IAEM	Institute of Higher Military Studies (Portugal)
IAF	Indian Air Force
IAFP	Inter-American Forces of Peace
IAFU	Improve Assault Fire Unit (USA)
IAFV	Infantry Armoured Fighting Vehicle
IAI	Israel Aircraft Industries Ltd
IAL	Infra-red Aiming Light
IALCE	International Airlift Control Element (NATO)
IAM	Inertially-Aided Munition
IAM	Initial Approach Mode
IANEC	Inter-American Nuclear Energy Committee
IA-PVO	Istrebitel 'naya Aviatsiya-PVO. Fighter aviation forces (CIS)
IARA	Inter-Allied Reparation Agency
IARRCIS	Interim ARRC Information System (NATO)
IAS	Improved Armour System
IAS	Indicated Air Speed
IAS	Integrated Acoustic Sensor
IASD	Instant Ammunition Selection Device
IAT	International Aerospace Technologies, Dr. Baumeister KG (Germany)
IATA	International Air Transport Association
IAU	ISDN Access Unit. Communications
IAW	Imagery Analysis Workpoint
IAW	Indications And Warning (NATO)
IB	Incendiary Bomb
IBC	International Business Consulting SA (Spain)
IBEC	International Bank for Economic Co-operation
IBERISA	Iberavia Internacional SA (Spain)

IBERLANT	Iberian Atlantic Command (NATO)
IBRD	International Bank for Reconstruction & Development
IBSR	Individual Battle Shooting Range (UK)
IBSS	Infra-red Background Sensor Survey
IC	In Charge
IC	Integrated Circuit
IC	Internal Combustion
ICA	International Communication Agency
ICAAS	Integrated Controls & Avionics for Air Superiority
ICAD	Individual Chemical Agent Detector
ICAO	International Civil Aviation Organisation
ICAR	Industria Condensatori Applicazioni Electtroelettroniche SpA (Italy)
ICBM	Inter-Continental Ballistic Missile
ICC	Information Co-ordination Centre
ICC	Information Co-ordination Control
ICC	Initial Connectivity Capability project (USA)
ICC	Integrated Communications Centre
ICCE	Iceland Command & Control Enhancement
ICCP	Improved Computer Control Panel
ICCS	Integrated / Improved Command & Control System
ICCS	IUKADGE Command & Control System
ICD	Interface Control Document
ICDO	International Civil Defence Organization (Switzerland)
ICDS	Integrated Control & Display System
ICDU	Integrated Control & Display Unit
ICE	Improved Combat Efficiency
ICE	Instant Camouflage Envelope
ICE	Interference Cancelling Equipment
ICE	Internal Communications Exchange
ICEROCC	Iceland Regional Operations Control Centre
ICJ	International Court of Justice
ICM	Improved Conventional Munition
ICM	Intercontinental Missile
ICM BB	Improved Conventional Munition, Base Bleed
ICMS	Integrated Countermeasures Suite
ICNIA	Integrated Communications Navigation & Identification Avionics
ICNICP	Integrated Communications / Navigation Identification Control Panel
ICNS	Integrated Communications / Navigation System
ICOM	Integrated COMSEC SINCGARS
ICON	Integration Contract
ICP	Incident Control Point
ICR	Intercooled Recuperated engine
ICRC	International Committee of the Red Cross
ICS	Improved Composite Structure
ICS	Integrated Communications System (UK)
ICS	Intercommunications System
ICS	Internal Countermeasures Set
ICSS	Improved Computer Sighting System
ICSS	Integrated Communications Switching System
ICSU	International Council of Scientific Unions

ICT	Fraunhofer-Institut fur Chemische Technologie (Germany)
ICU	Indicator Control Unit
ICU	Interface Converter Unit
ICV	Infantry Combat Vehicle
ICW	Interrupted Continuous Wave
ICWAR	Improved Continuous Wave Acquisition Radar
ID	Identification
ID	Intelligence Department
IDA	Installation Design Authority
IDA	Institute for Defense Analyses (USA)
IDA	International Development Association
IDACS	Integrated Digital Audio Control System
IDAS	Integrated Defensive Aids System
IDAS	Integrated Design Automation System
IDAS	International Danger & Disaster Assistance (Bahrain)
IDCSP	Initial Defense Communications Satellite Program (USA)
IDCT	Initial Detection & Classification Trainer
IDE	Interactive Development Environments Inc. (USA)
IDEA	International Defence Equipment & Aerospace Exhibition
IDEAL	Interactive Database Editor And Linker
IDEF	International Defence Industry & Civil Aviation Fair (Turkey)
IDEM	International Defence Market Show
IDEX	International Defence Exhibition
IDF	Iceland Defense Force (USA)
IDF	Indigenous Defence Fighter (Taiwan)
IDF	Instantaneous Direction Finder
IDF	Israeli Defence Forces
IDHS	Intelligence Data Handling System (USA)
IDL	International Data Link
IDLH	Immediately Dangerous to Life or Health
IDM	Impact Delay Module
IDR	International Development & Resources Inc. (USA)
IDS	Information Distribution & Switching system
IDS	Infra-Red Detection Set
IDS	Integrated Display Set (USAF)
IDS	Interdictor Strike
IDS	International Data Sciences Inc. (USA)
IDS	International Development Strategy
IDT	Identification & Tracking
IDVLS	Integrally-Ducted Vertical Launch System (UK)
IDW	Individual Defence Weapon
IEC	International Electrotechnical Commission
IED	Improvised Explosive Device
IED	Integrated Electric Drive (now known as ASMS) (USA)
IEE	Industrial Electronic Engineers Inc. (USA)
IEEE	Institute of Electrical & Electronic Engineers (USA)
IEMATS	Improved Emergency Message Automation Transmission System (USA)
IEMS	Installation Equipment Management System
IEPG	Independent European Programme Group (to be incorporated into the WEU)

IERC	International Electronic Research Co. (USA)
IERWITS	Initial Entry Rotary-Wing Integrated Training System (USA)
IES	Image Enhancement System
IEU	Interface Electronics Unit
IEW	Integrated Electronic Warfare
IEW	Intelligence / Electronic Warfare
IEWCS	Intelligence & Electronic Warfare Common Sensors
IF	Instrument Flying
IF	Intensive Flying (Royal Navy)
IFAB	Integrierte Feuerleitmittel fur Artillerie Batterie. Integrated fire-control system for artillery batteries (Germany)
IFANS	Institute of Foreign Affairs & National Security (Korea, South)
IFB	Invitation For Bid
IFCS	Improved Fire Control System
IFCS	Integrated Fire Control System
IFF	Identification Friend or Foe
IFI	Instruments for Industry Inc. (USA)
IFM	Instantaneous Frequency Measurement
IFMCS	Integrated Fire & Maneuver Control System
IFMU	Integrated Flight Management Unit
IFM/SHR	IFM Superheterodyne Receiver
IFOBL	In-Flight Operable Bomb Lock
IFR	Instrument Flight Rules
IFS	Indirect Fire Simulator
IFS	Inspectorate of Flight Safety (USAF, Royal Air Force)
IFT	Indirect Fire Trainer
IFT	Instructor Flown Target
IFT	International Flight Test (UK)
IFV	Infantry Fighting Vehicle
IFVwCM	Infantry Fighting Vehicle with Integrated Countermeasures (USA)
IFWES	Indirect Fire Weapon Effect Simulation
IF/FCS	Integrated Fire / Flight Control System
IG	Image Generator
IG	Instructor of Gunnery (Royal Air Force)
IGAC	Israeli General Avionics Computer
IGAT	Iranian Gas Trunkline
IGB	Inner German Border
IGC	Intergovernmental Conference
IGC	Interim Gunnery Computer
IGE	In Ground Effect
IGE	Instrumentation Graphics Environment
IGMDP	Integrated Guided Missile Development Programme (India)
IGN	Ignition
IGOR	Intercept Ground Optical Recorder
IGS	Internal Gun System
IHADSS	Integrated Helmet And Display Sighting System
IHAS	Integrated Helicopter Attack System
I-HAWK	Improved HAWK
IHE	Improved High Explosive

IHE	Insensitive High Explosive
IHEDN	Institut des Hautes Etudes de Defense Nationale (France)
IHFR	Improved HF Radio programme (USA)
IHI	Ishikawajima-Harima Heavy Industries Co. Ltd (Japan)
IHPI	Improved High Power Illumination
IHS	Integrated Headgear Sub-system
IHSC	International Headquarters & Support Command, SHAPE (NATO)
II	Image Intensification
II MEF	II Marine Expeditionary Force (USA)
IIDS	Integrated Instrumentation Display System
IIE GR	Institute of Industrial Exhibitions (Greece)
IIGF	Imperial Iranian Ground Forces
IIR	Imaging Infra-Red
IIR	Infra-red Imaging Radar (USA)
IIRA	Integrated Inertial Reference Assembly
IIRS	Instrument Inertial Reference Set
IIS	Infra-red Imaging System
IIS	Integrated Instrument System
IISA	Integrated Inertial Sensor Assembly
IISE	Intelligence Information Services Enhancement programme (USA)
IISS	International Institute for Strategic Studies (UK)
IITRI	IIT Research Institute (USA)
IJMS	Interim Joint Message System
IKAT	Interactive Keyboard And Terminal (Netherlands)
IKBS	Intelligent Knowledge-Based Systems
IKL	Ingenieurkontor Lubeck Prof. Gabler Nachf. GmbH (Germany)
IKW	Intercept & Kill Weapon
ILAAS	Integrated Low Altitude Attack Subsystem
ILAD	Inner Layer Air Defence system
ILAFA	Instituto Latinoamericano del Fierro y el Acero
ILG	International Logistics Group (USA)
ILI	Intamar Logistics Inc. (USA)
ILL	Illuminating
ILO	International Labour Organisation
ILS	Instrument Landing System
ILS	Integrated Logistic Support
ILSS	Integrated Logistic Support System
IMAAWS	Infantry Man-Portable Anti-armour / Assault Weapon System
IMAIS	Integrated Magnetic & Acoustic Influence Sweep
IMC	Instrument Meteorological Conditions
IMCO	Inter-Governmental Maritime Consultative Organisation
IMDEX	International Maritime Defence Exhibition
IME	Institute of Makers of Explosives
IME	Intercontinental Metals Export Co (USA)
IMET	International Military Education & Training
IMEWS	Integrated Missile Early Warning System
IMF	International Monetary Fund
IMFL	Institut de Mecanique des Fluides (France)
IMI	Israel Military Industries Ltd
IMICS	Integrated Mine-hunting Combat System

IMINT	Imagery Intelligence
IMK	Increased Manoeuvrability Kit
IMMLC	Improved Medium Mobility Load Class
IMO	International Maritime Organisation
IMO	International Meteorological Organisation
IMP	Infantry Mine Project
IMP	International Missile Proliferation
IMPAC	Integrated Message Processing And Communications system
IMPDS	Improved Missile Point Defence System
IMR	Inzhenernaia Maschina Razgrazhdeniia. Combat engineer vehicle (CIS)
IMS	Inertial Measurement System
IMS	Integrated Mission System
IMS	Integrated Multiplex System
IMS	International Marketing Services Ltd (USA)
IMS	International Military Services (UK)
IMS	International Military Staff (NATO)
IMS	Ion Mobility Spectrometer (Netherlands)
IMSE	Improved Mobile Subscriber Equipment
IMSS	Integrated Multi-Sensor System
IMT	International Military Tribunal
IMTS	Interactive Modular Training System
IMU	Inertial Measurement Unit
IN	Inertial Navigator
INAS	Integrated Navigation / Attack System
INC	Insertable Nuclear Component
INC	Ishikawajima Noise Control Co. Ltd (Japan)
INCA	Integrated Nuclear Communications Assessment (USA)
INCB	International Narcotics Control Board
INDEP	Industrias e Participacoes de Defesa (Portugal)
INESC	Institut National d'Etudes de la Securite Civile (France)
INEWS	Integrated Electronic Warfare System
inf	Infantry
INF	Intermediate-range Nuclear Forces (treaty)
INFOTERM	International Information Centre for Technology
INIS	International Nuclear Information System
INLA	Irish National Liberation Army
INMARSAT	International Maritime Satellite Oganisation
INRI	Inter-National Research Institute Inc. (USA)
INS	Inertial Navigation System
INSAS	Indian Small Arms System
INSCOM	Intelligence & Security Command (US Army)
INT	Intelligence
INTA	Instituto Nacional de Tecnica Aeroespacial (Spain)
INTACS	Integrated Tactical Communications Study (USA)
INTAF	Special Assistant to SACEUR for International Affairs (NATO)
INTAL	Instituto para la Integracion de America Latina
INTEL DIV	Intelligence Division, SHAPE (NATO)
INTELSAT	International Telecommunications Satellite Organization
INTERAMPOL	Inter-American Police

INTERATOMENEGO	International Economic Association for the Organization of Co-operation in Building Nuclear Power Stations
INTERATOMINSTRUMENT	International Economic Association for Nuclear Instrument Building
INTERCHIM	International Organization for Co-operation in Small-Tonnage Chemical Products
INTERCHIMVOLOKNO	International Economic Association for Chemical Fibres
INTERELEKTRO	International Organization for Co-operation in the Electrical Engineering Industry
INTERELEKTROTEST	International Organization for Co-operation in Large Capacity & High Voltage Experimental Laboratories
INTERMETAL	International Organization for Co-operation in the Iron & Steel Industry
INTERPAR	International Partnerships Group Inc (USA)
INTERPOL	International Criminal Police Organization
INTERSHIPNIK	International Organization for Co-operation of Bearings Industry
INTERSPUTNIK	International System & Organization for Space Communications
INTREP	Intelligence Report
INTSUM	Intelligence Summary
INU	Inertial Navigation Unit
IO	Identification Officer
IO	Intelligence Officer (UK)
IOC	Initial / Interim Operational Capability
IONDS	Integrated Operational Nuclear Detection System
IOR	Immediate Operational Requirement
IOS	Integrated Observation System
IOT	Integral Operator Trainer
IOT&E	Initial Operational Test & Evaluation
IOU	Interim OPCON Update (NATO)
IP	Initial Point
IP	Initial Provisioning
IP	Intermediate Pressure
IP	Intermediate Processor (communications)
IPADS	Improved Processing And Display System
IPAR	Improved Pulse Acquisition Radar
IPAT	Inertial Pointing Aided Tracking
IPCP	Improved Command Post
IPD	Improved Point Defence
IPD	Initial Production Delivery
IPD	Instituto de Pesquina e Desenvolvimento. Institute of research & development (Brazil)
IPDMS	Improved Point Defence Missile System
IPE	Image Processing Equipment
IPE	Individual Protective Equipment
IPE	Integral Protective Entrances
IPF	Initial Production Facility
IPI	Intercept Pattern for Identification
IPK	Individual Protection Kit (UK)

IPL	Information Processing Ltd (UK)
IPL	Instro Precision Ltd (UK)
IPLO	Irish People's Liberation Organization
IPM	Integrated Propulsion Module
IPM	Interpersonal Message
IPMS	Integrated Platform Management System
IPO	International Programme Office
IPP	International Precision Products BV (Netherlands)
IPqM	Instituto de Pesquisas da Marinas. Institute of Naval Research (Brazil)
IPSC	International Practical Shooting Confederation
IPSE	Integrated Project Support Environment
IPSS	Ice Penetrating Sensor System
IPT	Intelligent Procedure Trainer
IPTM	Institute of Pathology & Tropical Medicine. (Ministry of Defence)
IPTN	PT Industri Pesawat Terbang Nusantara (Indonesia)
IPTS	International Pistol Target System
IPU	Interface Processor Unit
IR	Infra-Red
IRA	Inertial Reference Assembly
IRA	Irish Republican Army
IRAN	Inspect & Repair As Necessary
I-RAP	Infra-Red Augmented Projectile
IRAS	Interdiction / Reconnaissance Attack System
IRAWS	Infra-Red Attack Weapon System
IRB	Improved Ribbon Bridge
IRB	Irish Republican Brotherhood
IRBM	Intermediate Range Ballistic Missile
IRCCD	Infra-Red Charge-Coupled Device
IRCCM	Infra-Red Counter Countermeasures
IRCM	Infra-Red Countermeasures
IRCS	Intrusion-Resistant Communications System
IRDF	Infra-Red Direction Finding
IRDS	Infra-Red Detecting Set
IRDU	Infra-Red Detection Unit
IREW	Infra-Red Electronic Warfare
IREWS	Infra-Red Electronic Warfare System
IRF	Immediate Reaction Force (NATO)
IRFIS	Inertial Referenced Flight Inspection System
IRFNA	Inhibited Red Fuming Nitric Acid (CIS)
IRI	Instituto per la Riconstruzione Industriali (Italy)
IRICS	Interim Reciprocal Information & Consultation System
IRIS	Infra-Red Imaging System
IRIS	Integrated Radar Imaging System
IRJ	Infra-Red Jammer
IRLS	Infra-Red Line Scanner
IRM	Inzhenernaya Rezvedivatel'naya Maschina. Engineer reconnaissance vehicle (CIS)
IRMS	Integrated Radio Management System
IRMWS	Infra-Red Missile Warning Subsystem programme (USA)
IROF	Improved Rate Of Fire

IRR	Infra-Red Radiation
IRR	Integral Rocket / Ramjet
IRR	Integrated Radio Room (USA)
IR-RAP	Infra-Red / Radar Augmented Projectile
IRROLA	Inflatable Radar-Reflective Optical Location Aid
IRRS	Infra-Red Reconnaissance System
IRS	Improved Radar Simulator
IRS	Inertial Reference System
IRSM	Infra-Red Surveillance Measures
IRSS	Infra-Red Search Set
IRST	Infra-Red Search & Track
IRSTD	Infra-Red Search & Target Designation
IRT	Infrared Technologies GmbH (Germany)
IRTH	Infra-Red Terminal Simulator
IRU	Inertial Reference Unit
IRV	Improved Recovery Vehicle
IRVAT	Infra-Red Video Automatic Tracking
IRWR	Infra-Red Warning Receiver
IR&D	Independent Research & Development
IR/EO	Infra-Red / Electro-Optical
IS	Intermediate System
IS	Internal Security
IS	International Staff, NATO Headquarters
ISA	Instruction Set Architecture
ISADS	Integrated Strapdown Air Data System
ISAR	Inverse Synthetic Aperture Radar
ISB	Independent Side Band
ISC	Infantry Section Carrier
ISC	Intercommunications Set Control
ISCOMFAR	Island Commander, The Faroes (NATO)
ISCOMICE	Island Commander, Iceland (NATO)
ISCS	Integrated Submarine Combat System
ISD	In-Service Date
ISD	Instructional Systems Development
ISDN	Integrated Services Digital Network
ISDS	Image Switching & Distribution System
ISDS	International Security & Defense Systems Ltd (Israel)
ISDS	IRCM Self-Defence System
ISE	International Submarine Engineering Ltd (Canada)
ISF	Integrated Support Facility
ISG	Information Strategies Group (USA)
ISG	Intelligence Support Group
ISI	Italia Sistemi Inerziali SpA
ISIS	Institute of Strategic & International Studies Malaysia
ISK	Intercept Sonar (Germany)
ISL	Institut de Saint-Louis (France)
ISLS	Interrogation Side-Lobe Suppression
ISN	Information Systems & Networks Corp. (USA)
ISO	International Standards Organisation
ISOR	Initial Statement of Requirement

ISP	Ideal Splash Point
ISPRA	Israel Product Research Co. Ltd
ISR	Identification – Safety Range
ISR	Image Storage & Retrieval
ISRP	Improved Stabilisation Reference Package
ISS	International Security Services Ltd (UK)
ISSAA	Information Systems Selection & Acquisition Activity office (US Army)
ISSR	Independent Secondary Surveillance Radar
ISST	ICBM Silo Superhardening Technology
IST	Invite, Show & Test
ISTARTLE	Integrated Surveillance Target Acquisition Radar for Tank Location & Engagement programme (USA)
ISTS	Information Systems (Technical Services)
ISU	In-arm Suspension Unit
ISU	Integrated Sight Unit
ISV	Internal Security Vehicle
ISVR	Institute of Sound & Vibration Research (UK)
ISVT	Integrated Secure Voice Terminal
ISYSCON	Integrated System Control programme (US Army)
IT	Intermediate Trainer
IT3	Interactive Tactical Team Trainer
ITAC	Intelligence Threat Analysis Center (USA)
ITACS	Integrated Tactical Command System
ITALD	Improved Tactical Air-Launched Decoy
ITAS	Improved Tracking Adjunct System
ITC	International Transducer Corp. (USA)
ITCM	Integrated Tactical Countermeasures (USAF)
ITCS	Integrated Target Control System
ITCS	Integrated Track & Control System
ITE	Involute Throat & Exit (rocket nozzle)
ITEC	International Turbine Engine Corp. (USA)
ITGS	Integrated Track Guidance System
ITH	Interturbine Holland BV
ITI	Industrial Tectonics Inc. (USA)
ITM	Industrial Technology & Machines AG (Switzerland)
ITO	International Trade Organisation
ITOW	Improved TOW missile
ITP	Intent To Purchase
ITP	Intention To Proceed
ITPS	International Test Pilots School (UK)
ITR	Integrated-Technology Rotor (US Army)
ITS	Initial Training Squadron (Royal Air Force)
ITS	Integrated Target System
ITS	Interactive Training System
ITS	International Thermal Sight
ITSA	International Technology SA (Spain)
ITSEC	Information Technology Security Evaluation Criteria
ITSEM	Information Technology Security Evaluation Method
ITSS	Integrated Tactical Surveillance System

ITT	Inter-Turbine Temperature
ITT	Invitation To Tender
ITU	International Telecommunications Union
ITV	Improved TOW Vehicle (USA)
ITW	Initial Training Wing (Royal Air Force)
IU	Interface Unit
IUKADGE	Improved United Kingdom Air Defence Ground Environment
IUS	Inertial Upper Stage
IUSS	Integrated Undersea Surveillance System
IVCS	Integrated Vehicle Communications System (USA)
IVCS	Integrated Voice Communication System
IVD	Interactive Video Disc
IVDN	Integrated Voice & Data Networking
IVDU	Intelligent Visual Display Unit
IVECO	Industrial Vehicle Corporation
IVIS	Inter-Vehicular Information System
IVIS	Inter-Vehicular Information System
IVMS	Integrated Vehicle Management System
IVPDL	Inter-Vehicle Positioning & Data Link (USA)
IVSI	Instantaneous VSI
IVSN	Initial Voice Switched Network (NATO)
IVV	Instantaneous Vertical Velocity
IW	Individual Weapon
IWAC	Integrated Weapon-Aiming System
IWC	Integrated Weapon Complex
IWD	Integrated Weapon Display
IWD	Interactive Weapon Display
IWESS	Infantry Weapons Effects Simulation System
IWS	Improved Weapon System
IWS	Individual Weapon Sight
IWS	Integrated Weapon System
IWSDB	Integrated Weapon System Data Base
IWT	Inzynieryjny Woz Torujacy. Combat engineer vehicle (Poland)
IX	Unclassified Miscellaneous ship (USA)
I/O	Input / Output

J

JAATT	Joint Air Attack Team Tactics
JAAWSC	Joint Anti-Air Warfare Shore Co-ordination Network (NATO)
JADDIN	Joint Air Defence Digital Information System (Thailand)
JADI	Japan Association of Defence Industries
JAE	Japan Aviation Electronics Industry Ltd
JAFE	Joint Advanced Fighter Engine (USA)
JAGUAR-V	Jamming Guarded radio. VHF frequency hopping radio system (UK)
JAIEG	Joint Atomic Information Exchange Group
JAMCAT	Jammer Communications Attachment
JANAP	Joint Army, Navy, Air Force Publication
JAPNMS	JTIDS Air Platform Network Management System
JAR	Joint Airworthiness Requirements
JARIC	Joint Air Reconnaissance Centre (UK)
JASDF	Japan Air Self Defence Force
JASORS	Joint Advanced Special Operations Radio System (USA)
JASS	Joint Anti-Submarine School (UK)
JATCCCS	Joint Advanced Tactical C3 System (USA)
JATO	Jet Assisted Take-Off
JATS	Jamming Analysis & Transmission Selection
JAWS	Jamming And Warning System
JAWS	Joint Attack Weapon Systems (USA)
JBUSDC	Joint Brazil-US Defense Commission
JCALS	Joint Computer-aided Acquisition & Logistic Support system
J-CATCH	Joint Countering Attack Helicopters program (USA)
JCIC	Joint Compliance & Inspection Committee
JCMPO	Joint Cruise Missile Project Office (USAF & Navy)
JCS	Joint Chiefs of Staff
JDA	Japanese Defense Agency
JDAM	Joint Direct Attack Munition programme (USA)
JDCU	Jamming Detection Control Unit
JDF	Jamaica Defence Force
JDFCG	Jamaica Defence Force Coast Guard
Jengo	Junior Engineering Officer (Royal Air Force)
JESS	Joint Exercise Support System
JETDS	Joint Electronics Type Designation System (NATO)
JEWC	Joint Electronic Warfare Center (USA)
JFDP	Joint Force Development Process (USA)
JFHQ	Joint Forces Headquarters
JGSDF	Japan Ground Self-Defence Force
JHP	Jacketed Hollow Point
JHQ	Joint Headquarters
JHSU	Joint Helicopter Support Unit (UK)
JIAWG	Joint Integrated Avionics Working Group (USA)
JIES	Joint Interoperability Evaluation System (USA)
JIFDATS	Joint-services In-Flight Data Transmission System (USA)
JIG	Jane's Information Group Ltd (UK)
JIMPACS	Joint Improved Multi-mission Payload Aerial Surveillance, Combat Survivable (now Hunter)

JINTACS	Joint Interoperability of Tactical Command And Control System (USA)
JIP	Joint Interface Program (USA)
JIT	Just-In-Time
JITA	Japan Industrial Technology Association
JJPTP	Joint Jet-Pilot Training Programme
JMC	Joint Maritime Course
JMEA	Japan Machinery Exporters' Association
JMEM	Joint Munitions Effectiveness Manual (USA)
JMIC	Joint Maritime Intelligence Centre
JMOTS	Joint Maritime Operational Training Staffs
JMSA	Japan Maritime Safety Agency
JMSDF	Japan Maritime Self-Defence Force
JMUSDC	Joint Mexican-US Defense Commission
JNA	Former Yugoslav National Army
JOC	Joint Operations Centre
JOG	Joint Operations Group (USA)
Joint STARS	Joint Surveillance & Target Attack Radar System (also JSTARS) (USA)
JOP	Junior Officer Pilot (Royal Air Force)
JORN	Jindalee OTH Radar Network (Australia)
JOSIC	Joint Ocean Surveillance Information Centre
JOTS	Joint Operational Tactical System
JP	Jamming Pulse
JPATS	Joint Primary Aircraft Training System (USA)
JPI	Joint Precision Interdiction. New NATO doctrine to succeed FOFA
JPL	Jet Propulsion Laboratory, NASA (USA)
JPO	Joint Program Office (USA)
JPTT	Junior Officer Tactics Team (US Navy)
JRA	Jaguar Rover Australia
JRA	Japanese Red Army
JRD-3	Japan Reconfiguration & Digitisation Program Phase III (USA)
JRI	Jules Richard Instruments (France)
JRSC	Jam-Resistant Secure Communications programme
JSC	Johnson Space Center. NASA (USA)
JSC	Joint Steering Committee
JSCMPO	Joint Service Cruise Missile Program Office (USAF & Navy)
JSDC	Joint Service Defence College (UK)
JSDF	Japan Self Defence Force
JSDFA	Japan Self Defence Force Agency
JSESPO	Joint Surface-Effect Ship Program Office (USA)
JSFU	Joint Services Flail Unit
JSIPS	Joint Service Image Processing System
JSOA	Joint Special Operations Agency (USA)
JSOR	Joint Services Operational Requirement
JSOW	Joint Service Stand-Off Weapon (USA)
JSP	Jacketed Soft Point
JSP	Joint Services Publication
JSR	Jammer Saturation Range
JSRJ	Joint Services Recognition Journal

JSRR	Jam / Signal Ratio Required
JSS	Joint Surveillance System
JSSC	Joint Services Staff College (Australia)
JSSG	Joint Signal Support Group (NATO)
JSTARS	Joint Surveillance & Target Attack Radar System (USA)
JSWDL	Joint Service Weapon Data Link (USA)
JTACMS	Joint Tactical Missile System
JTC3A	Joint Tactical Command, Control & Communications Agency (USA)
JTF	Joint Task Force
JTFP	Joint Tactical Fusion Programme (USA)
JTIDS	Joint Tactical Information Distribution System (USA)
JTMMR	Joint Tactical Multi-Mode Radio (USA)
JUME	Junta de Metodos de Tir (Spain)
JUMP	Joint UHF Modernisation Project (USA)
JV	Joint Venture
JVAR	Jordan Valley Applied Radiation Ltd (Israel)
JVX	Joint service advanced vertical lift aircraft (USA)
J/S	Jam-to-Signal ratio

K

KADS	Knowledge Acquisition Data System
KAS	Killed on Active Service
KATIE	Killer Alert Threat Identification & Evasion
KBSL	Knowledge Base Services Ltd (UK)
KBU	Keyboard Unit
KC	The Kent Chemical Co. Ltd (UK)
KCl	Potassium chloride
KCS	Kockums Computer Systems AB (Sweden)
KD	Known Distances
KDAR	Kleindrohne Anti-Radar. Harrassment drone (Germany)
KDB	Kapal Di-Raja Brunei
KDIA	Korea Defense Industry Association (Korea, South)
KdoMFuSys	Kommando Marine Fuhrungssystems (Germany)
KDR	Kill / Detection Ratio
KE	Kinetic Energy
KEC	Korea Explosives Co. Ltd
KEK	Kinetic Energy Kill
KEM	Kinetic Energy Missile (USA)
KEP	Kinetic Energy Penetrator
KERR	Kinetic Energy Recovery Rope
KEW	Kinetic Energy Weapon
KFP	Korean Fighter Program (Korea, South)
KGB	Committee for State Security (CIS)
KH	Kelvin Hughes A/S (Denmark)
KhAB	Khimicheskayo AviaBomba. Chemical air-launched bomb (CIS)
KHD	Klockner-Humboldt-Deutz AG (Germany)
KHI	Kawasaki Heavy Industries Ltd (Japan)
KIA	Killed In Action
KIAS	Knots Indicated Air Speed
KITE	Kinetic energy kill vehicle Integrated Technology Experiments
KKMC	King Khalid Military City (Saudi Arabia)
KKV	Kinetic Kill Vehicle
KKW	Kinetic Kill Weapon
KMA	Korea Military Academy (Korea, South)
KMA	Royal Military Academy for Army & Air Force (Netherlands)
KOSI	Kaiser Optical Systems Inc. (USA)
KP	Key Point
KPAF	Korean People's Air Force (Korea, North)
KPNLF	Khmer People's National Liberation Front
KPS	Knowledge Processing System
KPU	Keyboard Printer Unit
KSC	Kennedy Space Center (USA)
KSE	Kewdale Structural Engineers (WA) (Australia)
KShM	Komandno-Shtabnaya Mashina. Command-staff vehicle (CIS)
KSR	Keyboard Send / Receive
KTC	Korea Technologies Corp. (Korea, South)
KVDT	Keyboard Visual Display Unit
KWIC	Key Word In Context
KWOC	Key Word Out of Context

L

L	Lewisite. Chemical warfare blister agent
L3TV	Low Light Level TV
LAA	Light Anti-Aircraft
LAAAS	Low-Altitude Airfield Attack System
LAAD	Low Altitude Air Defence
LAADS	Low Altitude Aircraft Detection System
LAAG	Light Anti-Aircraft Gun
LAAM	Light Anti-Aircraft Missile
LAASH	LITEF's Analogue Air data System for Helicopters
LAAT	Laser Augmented Airborne TOW
LAAV	Light Airborne ASW Vehicle
LAB	Light Assault Bridge
LABCOM	Laboratory Command (US Army)
LABRV	Large Advanced Ballistic Re-entry Vehicle
LABS	Low-Altitude Bombing System
LAC	Leading Aircraft(s)man
LACE	Landline Air defense Communications Encryption (USA)
LACH	Lightweight Amphibious Container Handler
LACI	Lockheed Aeromod Centers Inc. (USA)
LACW	Leading Aircraftwoman (Royal Air Force)
LAD	Large Area Display
LAD	Laser Acquisition Device
LAD	Launch Assist Device
LAD	Leurre Actif Decale. Common term for offboard decoy systems
LAD	Liberation Army Daily (newspaper) (China)
LAD	Light Aid Detachment
LAD	Low Altitude Dispenser
LADAR	Laser Detection And Ranging
LADD	Low Altitude Drogue Delivery
LADS	Laser Airborne Depth Sounder
LADS	Light Air Defense System (USA)
LAE	Low Altitude Extraction
LAF	Light Assault Ferry
LAF	Low Frequency Active System
LAFTA	Latin American Free Trade Association
LAFTS	Laser And FLIR Test Set
LAH	Light Attack Helicopter
LAIA	Latin American Integration Association
LAIR	Laser & Atomic Research & Development (Croatia)
LAIRS	Light Aircraft Reconnaissance System
LAL	Launch And Leave
LAM	Long Aerial Mine
LAMP	Large Advanced Mirror Programme
LAMP	Lockheed Adaptive Modular Payload
LAMPS	Light Airborne Multi-Purpose System
LAMS	Local Area Missile System
LAN	Local Area Network

LANA	Low Altitude Night Attack
LANDCENT	Allied Land Forces, Central Region (NATO)
LANDJUT	Allied Land Forces, Schleswig-Holstein & Jutland (NATO)
LANDNOR	Allied Land Forces, North Norway (NATO)
LANDSOUTH	Allied Land Forces, Southern Europe (NATO)
LANDSOUTHEAST	Allied Land Forces, Southeastern Europe (NATO)
LANDZEALAND	Allied Land Forces, Zealand (NATO)
LANL	Los Alamos National Laboratory (USA)
LANTIRN	Low Altitude Navigation & Targeting Infra-Red for Night
LAP	London Artid Plastics Ltd (UK)
LAPA	Luftgetutzes Abstandsfahiges Primar Aufklarungssystem (Germany)
LAPADS	Lightweight Acoustic Processing And Display System
LAPES	Low Altitude Parachute Extraction System
LAPTC	Lucas Aerospace Power Transmission Corp. (USA)
LAR	Light Artillery Rocket
LAR	Light Automatic Rifle
LaRC	Langley Research Center (USA)
LARC	Lighter Amphibious Resupply Cargo
LARC	Low Altitude Ride Control
LARCEF	Latin American Council for Cosmic Radiation & Physical Interplanetary Space
LARS	Light Artillery Rocket System
LAS	Lockheed Aircraft Services Co. (USA)
LASC	Lockheed Aeronautical Systems Co. (USA)
LASER	Light Amplification by Stimulated Emission of Radiation
LASHE	Low Altitude Simultaneous HAWK Engagement
LASO	Latin American Solidarity Organization
LASR	Low Altitude Surveillance Radar
LASS	Large Area Smoke Screening
LASS	Local Area Sensor System (UK)
LASS	Low Altitude Surveillance System
LASSIE	Low Airspeed Sensing & Indicating Equipment
LASSO	Light Anti-Surface Semi-automatic Optical missile (France)
LAST	Light Applique System Technique
LAST	Low Altitude Supersonic Target
LASTE	Low Altitude Safety & Targeting Enhancement
Lat	Latitude
LAT	Liebherr-Aero-Technik GmbH (Germany)
LAT	Light Anti-Tank
LATAR	Laser-Augmented Target Acquisition & Recognition
LATAS	Laser True Airspeed System
LATCC	London Air Traffic Control Centre (UK)
LATEX	Laser Associe a une Tourelle Experimentale. Laser system in experimental turret (France)
LATIS	Lightweight Airborne Thermal Imaging System
LATS	Light Armoured Turret System
LAV	Light Armored Vehicle (USA)
LAV	Light Assault Vehicle (USA)
LAVA	Low-profile Adaptive Vehicular Antenna

LAV-AD	Light Armoured Vehicle – Air Defence (USA)
LAV-AG	Light Armoured Vehicle – Assault Gun
LAW	Light Anti-armour Weapon
LAW	Light Anti-tank Weapon
LAX	Limited Area automatic Extraction (radar tracking)
LB	Light Bomber
LB	Local Battery
LBA	Luftfahrtbundesamt. Civil Aviation Authority (Germany)
LBH	Light Battlefield Helicopter
LBR	Laser Beam Rider
LBRG	Laser Beam-Riding Guidance
LBTS	Land-Based Test System
LCA	Light Combat Aircraft
LCAC	Landing Craft, Air Cushion (USA)
LCAJ	Low-Cost Anti-Jam system
LCAN	Lockheed Canada Inc.
LCAS	Light Close Air Support
LCC	Amphibious Command Ship (USA)
LCC	Launch Control Centre
LCC	Life Cycle Cost
LCC	Limited Capability Configuration
LCC	Local Command Centre
LCC	Logistics Co-ordination Centre
LCCC	Launch Control Centre Computer
LCD	Liquid Crystal Display
LCdr	Lieutenant-Commander
LCEE	Low Cost Emplacement Excavator
LCF	Le Cablage Francais
LCF	Low Cycle Fatigue
LCH	Light Combat Helicopter
LCINS	Low Cost Inertial Navigation System
LCISA	Lockheed Corporation International SA (Switzerland)
LCLU	Landing Control Logic Unit
LCM	Landing Craft, Mechanised (USA)
LCM	Laser Countermeasures
LCN	Load Classification Number
LCO	Launch Control Officer
LCOSS	Lead Computing Optical Sight System
LCP	Landing Craft, Personnel (USA)
LCP	Launch Control Post
LCpl	Lance Corporal
LCR	Laboratoire Central de Recherches, Thomson-CSF (France)
LCS	Low-Cost Sonobuoy
LCSS	Land Combat Support System
LCT	Landing Craft, Tank
LCT	Light Crawler Tractor
LCU	Landing Craft, Utility (USA)
LCU	Launch Control Unit
LCU	Lightweight Computer Unit
LCV	Low-Cost Visual

LCVP	Landing Craft, Vehicle, Personnel (USA)
LCWDS	Low Cost Weapon Delivery System
LD	Loop Disconnect
LD	Low Drag (bombs)
LDC	Launch Detection System
LDDI	Less Developed Defence Industrial nation
LDF	Lightweight Digital Facsimile
LDNS	Laser Doppler Navigation System
LDO	Limited Duties Officer (US Navy)
LDOCF	Long Distance Operational Control Facility
LDP	Laser Designator Pod
LDR	Low Data Rate
LDS	Laser Dazzle Sight
LDS	Layered Defence System
LDT	Laser Detector Tracker
LDTM	Light-Duty Tank Target Mechanism
LDT/SCAM	Laser Detector & Tracker-Strike Camera
LDU	Launcher Display Unit
LEAP	Lightweight Exoatmospheric Advanced Projectile
LECOS	Light Electronic Control System
LED	Light Emitting Diode
LED	Low Endoatmospheric Defence
LEDA	Laser Electrique a Detente Adiabatique. Gas-dynamic laser (France)
LEDI	Low Endoatmospheric Defence Interceptor
LEGS	Lightweight Engine Generator Set
LEN	Large Extension Node. Communications
LENS	Large Extension Node Switch. Communications
LEO	Low Earth Orbit
LEOS	Loral Electro-Optical Systems
LEP	Light External Pintle (Belgium)
LER	Laboratoire Electroniques de Rennes, Thomson-CSF (France)
LER	Life Extension Refit
LeRC	Lewis Research Center (USA)
LESC	Lockheed Engineering & Sciences Co. (USA)
LET	Launch & Escape Time
LETRI	Leihua Electronic Technology Research Institute (China)
LEU	Laser Electronic Unit
LEW	Lyttleton Engineering Works
LEWDD	Lightweight Early Warning Detection Device
LF	Landing Force
LF	Launch Facility
LFA	Low Frequency Active sensor
LFH	Low Fire Hazard
LFICS	Landing Force Integrated Communications System
LFM	Live Firing Monitor
LFP	Light Floor Pintle (Belgium)
LFRED	Liquid-Fuelled Ramjet Engine Development
LFRJ	Liquid-Fuelled Ramjet
LFSMS	Logistic Force-Structure Management System
LFV	Light Forces Vehicle

LFX	Limited output Full area automatic Extraction (radar tracking)
LGB	Laser-Guided Bomb
LGDM	Laser-Guided Dispenser Munition
LGM	Loop Group Multiplexer. Communications
LGM	Silo-Launched surface attack Guided Missile
LGR	Laser Guidance Receiver
LGSC	Linear Glideslope Capture
LH	Light Helicopter
LHA	Amphibious Assault Ship, general purpose (USA)
LHD	Amphibious Assault Ship, multi-purpose (USA)
LHD	Left-Hand Drive
LHM	Laser-Hardened Materials
LHN	Long-Haul Network
LHS	Load Handling System
LHTEC	Light Helicopter Turbine Engine Co. (USA)
LHW	Laser-Homing Weapon
LHX	Light Helicopter Experimental (US Army)
LI	Laser Interrogator
LIA	Linear Induction Accelerator
LIB	Loudspeaker Intercom Box
LIC	Line Integrity Check
LICAS	Low Intensity Conflict Aircraft System
LID	Laser Irradiation Detector
LID	Lift-Improvement Device
LIDAR	Light Detection And Ranging
Liet	Lieutenant
LIFT	Lead-In Fighter Training
LIG	Laser Image Generator
LII	Light Image Intesifier
LIM	Low-Inclination Mission
LIMA	Langkawi International Maritime & Aerospace Exhibition (Malaysia)
LIMAS	Lightweight Marking System
LIMSS	Logistics Information Management Support System
LIN	Linjeflyg AB (Sweden)
LINAC	Linear Accelerator
LINAS	Laser-Inertial Navigation / Attack System
LINS	Laser Inertial Navigation System
LIOD	Lightweight Optronic Director (Netherlands)
LIR	Laser Intercept Receiver
LIRA	Low Intesity Reconnaissance Aircraft
LIRD	Laser & Infra-red Irradiation Detector (Yugoslavia)
LIROD	Lightweight Radar Optronic Director (Netherlands)
LIRS	Laser Inertial Reference System
LISE	Laser Intergrated Space Experiment (USA)
LITACS	Lightweight Integrated Tactical Artillery Command & Control System
LITAL	Litton Italia SpA
LITBIEL	Lithuania & Byelorussia
LITDL	Link-16 Interoperable Tactical Data Link (USA)
LITE	Laser Illuminator Target Equipment
LITKOR	Litton Korea Ltd

LITS	Logistics Information Technology Strategy (Royal Air Force)
LITVC	Liquid-Injection Thrust Vector Control
LIVEX	Live Exercise (NATO)
LIVH	Liaison Intervehicule Hertzienne. Vehicle radio link & mast (France)
LIW	Lyttleton Engineering Works (South Africa)
LKA	Amphibious Cargo Ship (USA)
LKL	Lochkegelleitwerk (Germany)
LL	Low Level
LLAD	Low Level Air Defence
LLGB	Launch & Leave Guided Bomb
LLGB	Low Level Guided Bomb
LLH	Light Liaison Helicopter
LLLGB	Low-Level Laser Guided Bomb
LLLTV	Low Light Level Television
LLNL	Lawrence Livermore National Laboratory (USA)
LLTV	Low Light Television
LLWD	Low-Level Weapons Delivery
LMA	Leading Medical Assistant (Royal Navy)
LMAW	Light Multi-purpose Assault Weapon
LMCR	Liquid Metal Cooled Reactor
LMD	Light Mobile Digger
LMDE	Limpet Mine Disposal Equipment
LMG	Light Machine Gun
LMI	Le Magnesium Industriel (France)
LML	Lightweight Multiple Launcher
LML (N)	Lightweight Multiple Launcher (Naval)
LML (V)	Lightweight Multiple Launcher (Vehicle)
LMMSA	Lightweight Modular Multi-purpose Spanning Assembly
LMOTS	Launcher Maintenance & Operational Training System
LMRDFS	Lightweight Man-transportable Radio Direction Finding System
LMRS	London Military Radar Services
LMSC	Lockheed Missiles & Space Co. Inc. (USA)
LMTR	Laser Marker & Target Ranger
LMTV	Light Medium Tactical Vehicle
LMTVT	Light Medium Tactical Vehicle Trailer
LNA	Low Noise Amplifier
LNE	Late Network Entry
LNO	Limited Nuclear Option
LNS	Land Navigation System
LNTS	League of Nations Treaty Series
LO	Liaison Officer
LO	Local Oscillator
LO	Low Observables (UAV)
LOA	Launch On Assessment
LOAD	Low Altitude Defence
LOADEO	Loading of Explosive Ordnance
LOADS	Low-Altitude Air Defence System
LOAL	Lock-On After Launch
LOB	Line-Of-Bearing
LOBL	Lock-On Before Launch

LOC	Lines Of Communication
LOC	Local Operations Console
LOC	Localiser (element of ILS)
LOC	Logistics Operations Centre
LOCAAS	Low-Cost Anti-Armor Submunition programme (USA)
LOCAP	Lo Combat Air Patrol
LOCAT	Low Cost Artillery Trainer
LOCAT	Low-Cost Aerial Trainer
LOCATM	Low-Cost Advanced Technology Missile
LOCC	Launch Operations Control Centre
LOCE	Large Optical Communications Experiment
LOCE	Limited Operational Capacity for Europe
LOCLAD	Low-Cost Low Altitude Dispenser (USA)
LOCPOD	Low Cost Powered Off-boresight Dispenser
LOCREP	Location Report
LOCUS	Laser Obstacle Cable Unmasking System
LODE	Large Optics Demonstration Experiment
LOE	Level Of Effort
LOE	Limit Of Exploitation
LOF	Lift-Off
LOF	Line-Of-Fire
LOFAADS	Lo-Altitude Forward Area Air Defense System (USA)
LOFAR	Low Frequency Omnidirectional Acoustic Frequency Analysis & Recording
LOFARGRAM	Low Frequency Recording & Analysis Gram
LOGHOLDAIR	Air Logistics Message (NATO)
LOGMAN	Logistics & Manpower Division, SHAPE (NATO)
LOGSHORE	Logistics Short Report message (NATO)
LOGSUM	Logistics Summary message (NATO)
LOGSUPREP	Logistics Support Report message (NATO)
LOH	Light Observation Helicopter
LOMADS	Low Altitude Missile Air Defence System
LOMAH	Location of Miss And Hit
long	Longitude
lopro	Low probe. Military aircraft mission
LOPT	Lynx Helicopter Observer Procedure Trainer
LORAAS	Long-Range Airborne ASW System
LORAN	Long Range Aid to Navigation
LORIS	Long-Range Infra-red System
LORO	Lobe On Receive Only
LOROC	Long-Range Offboard Chaff
LOROC	Long-Range Optical Camera
LOROP	Long-Range Oblique Photography
LORROS	Long-Range Reconnaissance & Observation System
LORV	Low-Observable Re-entry Vehicle
LOS	Line-Of-Sight
LOSACA	Liaison Officer to the Supreme Allied Commander, Atlantic (USA)
LOSAT	Line-Of-Sight Anti-Tank
LOS-F	Line-Of-Sight Forward
LOSI	Line-Of-Sight Indicator

LOS-R	Line-Of-Sight Rear
LOTAWS	Laser Obstacle Terrain-Avoidance Warning System
LOTS	Logistics Over The Shore
LOVA	Low-Vulnerability Ammunition
LOW	Launch On Warning
LOX	Liquid Oxygen
LP	Liquid Propellant
LP	Listening Post
LP	Low Pressure
LPA	Amphibious Transport (USA)
LPA	Linear Power Amplifier
LPAR	Large Phased Array Radar
LPC	Launch Pod Container
LPC	Linear Predictive Coding
LPC	Low Pressure Compressor
LPD	Amphibious Transport Dock (USA)
LPD	Label Plan Display
LPD	Labelled Position Display
LPH	Amphibious assault ship, helicopter (USA)
LPH	Light Pintle Head (Belgium)
LPI	Low Probability of Interception
LPIR	LPI Radar
LPI/LPE	Low Probability of Interception & Exploitation
LPL	Linear Polarised Laser
LPM	Lines Per Minute
LPR	Low-cost Packet Radio
LPS	Lhotellier Plastiques Sologne SA (France)
LPT	Low Pressure Turbine
LPT	Low Profile Turret
LPTS	Lightweight Protected Turret System
LPU	Local Programming Unit
LP/C	Launch Pod / Container
LQA	Link Quality Analysis
LR	Long Range
LRA	Soviet Long-Range Aviation
LRAACA	Long-Range Air Anti-submarine Capability Aircraft
LRAAS	Long-Range Airborne ASW System
LRAS3	Long-Range Advanced Scout Surveillance System
LRAT	Large Radar Array Technology
LRAT	Long-Range Anti-Tank
LRATGW	Long-Range Anti-Tank Guided Weapon
LRBA	Laboratoire de Recherches Balistiques et Aerodynamiques (France)
LRBB	Long-Range, Base Bleed
LRBM	Long-Range Ballistic Missile
LRCA	Long-Range Combat Aircraft
LRCR	Long-Range Chaff Rocket
LRCS	Long-Range Communications System
LRCSOW	Long-Range Coventional Stand-Off Weapon
LRCU	Landing Rollout Control Unit
LRD	Labelled Radar Display

LRD	Laser Rangefinder / Designator
LRF	Laser Rangefinder
LRF	Low Recoil Force
LRF/D	Laser Rangefinder / Designator
LRGB	Long-Range Glide Bomb
LRHB	Long-Range, Hollow Base
LRINF	Longer-Range Intermediate Nuclear Forces
LRIP	Low Rate Initial Production
LR-IST	Long-Range Infra-Red Search & Track
LRL	Lightweight Rocket Launcher
LRM	Line Replaceable Module
LRM	Long-Range air-to-air Missile
LRM	Low Rate Multiplexer
LRMP	Long Range Maritime Patrol
LRMR	Long Range Maritime Reconnaissance
LRMTS	Laser Ranger & Marked Target Seeker
LRN	Low Recoil Noricum
LRPA	Long-Range Patrol Aircraft
LRRP	Long Range Reconnaissance Patrol
LRSAM	Long Range Surface-to-Air Missile
LRSOM	Long-Range Stand-Off Missile
LRSS	Long-Range Surveillance System
LRTNF	Long-Range Theatre Nuclear Forces
LRU	Line Replaceable Unit
LRV	Launch & Recovery Vehicle
LRV	Light Rail Vehicle
LS	Leading Seaman
LSA	Logistics Support Analysis
LSAT	Logistic Shelter, Air-Transportable
LSB	Lower Sideband
LSCP	Launching System Control Panel
LSD	Landing Ship, Dock (USA)
LSD	Large Screen Display
LSD	Lysergic Acid Diethylamide
LSDIS	Light & Special Division Interim Sensor
LSE	Local Security Environment
LSFFAR	Low-Speed Folding-Fin Aircraft Rocket
LSI	Landing Ship, Infantry
LSI	Large Scale Integration
LSI	Life Support International Inc. (USA)
LSIC	Large Scale Integrated Circuit
LSL	Landing Ship, Logistic
LSM	Landing Ship, Medium
LSO	Limited Strategic Option
LSOC	Lockheed Space Operations Co. (USA)
LSOT	Landing Signal Officer Trainer
LSP	Logistics Support Plan
LSSC	Light Seal Support Craft (USA)
LSSI	Lockheed Support Systems Inc. (USA)
LSST	Laser Spot Seeker / Tracker

LST	Landing Ship, Tank (USA)
LST	Laser Spot Tracker
LSU	Line Signalling Unit
LSV	Light Strike Vehicle
LSV	Logistic Support Vehicle
LSVW	Logistic Support Vehicle, Wheeled
LSW	Light Support Weapon
Lt	Lieutenant
LT	Light Transportable
Lt Cdr	Lieutenant Commander
Lt jg	Lieutenant, Junior Grade (US Navy)
LTA	Light Transport Aircraft
LTBT	Limited Test Ban Treaty
LTC	Long Term Contract
Lt-Col	Lieutenant-Colonel
LTD	Laser Target Designator
LTDP	Long-Term Defence Programme
LTDS	Laser Target Designator System
LTD/R	Laser Target Designator / Ranger
LTFCS	Laser Tank Fire Control System
LTFP	Long Term Force Programmes (NATO)
LTG	Line Termination Group
Lt-Gen	Lieutenant-General
LTH	Light Towed Howitzer
LTI	Laser Technology Inc. (USA)
LTID	Laser Target Interface Device
LTM	Laser Target Marker
LTM	Line Termination Module
LTMR	Laser Target Marker & Receiver
LTOC	Lockheed Technical Operations Co. (USA)
LTR	Loop Transfer Recovery
LTS	Laser Training System
LTS	Link Translator System
LTS	Low frequency Transmit Subsystem
LTT	Light Tracked Tractor
LTU	Laser Transceiver Unit
LTV	Lhotellier Repack Technische Verpackungen GmbH (Germany)
LTWA	Long Trailing Wire Antenna
LUA	Launch Under Attack
LUNOS	Lightweight Universal Night Observation System
LVA	Landing Vehicle, Assault (USA)
LVA	Large Vertical Aperture radar
LVDT	Linear Variable Differential Transformer
LVS	Logistic Vehicle System
LVT	Landing Vehicle, Tracked (USA)
LVTC	Landing Vehicle, Tracked, Command (USA)
LVTE	Landing Vehicle, Tracked, Engineer (USA)
LVTH	Landing Vehicle, Tracked, Howitzer (USA)
LVTP	Landing Vehicle, Tracked, Personnel (USA)
LVTR	Landing Vehicle, Tracked, Recovery (USA)

LW	Long Wave
LWA	Laser Warning Analyser
LWB	Long Wheelbase
LWD	Laser Warning device
LWDR	Lightweight Designator / Rangefinder
LWF	Long-Weight Fighter
LWIR	Long Wave Infra-Red
LWL	Lightweight Weapon Launch system
LWLD	Lightweight Laser Designator
LWML	Light Weight Multiple Launcher
LWMS	Light Weight Modular Thermal Sight
LWP	Light Window Pintle (Belgium)
LWR	Laser Warning Receiver
LWT	Amphibious Warping Tug (USA)
LWT	Light Weight Turret
LZ	Landing Zone
LZE	Luminous-Zone Emissivity
L&TH	Lethality & Target Hardening
L/O	Lift-Off

M

M2F2	Multimode Fire & Forget
MA	Medical Assistant (Royal Navy)
MAA	Master At Arms (Royal Navy)
MAA	Mission Area Analysis
MAA	Monitoring Angle of Attack
MAB	Marine Amphibious Brigade (USA)
MAB	Mobile Assault Bridge
MAC	Mandatory Access Control
MAC	Mean Aerodynamic Chord
MAC	Medium Armored Car (USA)
MAC	Military Airlift Command (USAF)
MACC	Military Area Control Centre
MACCS	Marine Air Command Control System (USA)
MACCS	Mobile Adaptable Communications Countermeasures System
MACEN	Air Defense Command & First Air Region (Spain)
MACHAN	Mando Aereo de Canarias (Spain)
MACI	Military Adaption of Commercial Items
MACI	Mine Anti-Char Indetectable (France)
MACIMS	Military Airlift Command Integrated Management System
MACOM	Mando Aero de Combate. Air defence system (Spain)
MACS	Magnetic Countermine System
MACS	Marine Air Control Squadron (USA)
MACS	Multiple Acceleration Control System
MACS	Multiple Access Communications System
MACS	Multiple Applications Control System
MACTIS	Mine-hunting Action Information Subsystem
MAD	Magnetic Anomaly Detector
MAD	Mass Air Delivery
MAD	Mutually Assured Destruction
MADAR	Maintenance Analysis Detection And Recording
MADC	Miniature Air Data Computer
MADGE	Malaysian Air Defence Ground Environment
MADGE	Microwave Aircraft Digital Guidance Equipment
MADLS	Mobile Air Defence Launching System
MADS	Military Advanced Disk System (computer)
MADS	MPS Air Defence Simulator
MAE	Marine & Aerospace Engineering Pty Ltd (Australia)
MAE	Mean Area of Effectiveness (USAF)
MAEO	Master Air Electronics Officer (Royal Air Force)
MAEST	Mando Aereo del Estrecho (Spain)
MAF	Malaysian Armed Forces
MAF	Manual Acquisition Facility
MAF	Marine Amphibious Force (USA)
MAF	Material Amphibie de Franchissement (France)
MAFLIR	Modified Advanced Forward Looking Infra-Red
MAG	Marine Air Group
MAG	Military Assistance Group
MAG	Mitrailleuse a Gaz. Machine gun (Belgium)

MAGICS	Modular Architecture for Graphics & Image Control System
MAGIK	Merit Automated Graphics Interface Kit for simulation system
MAGLAD	Marksman And Gunnery Laser Device
MAGR	Miniature Airborne GPS Receiver
MAGTF	Marine Air Ground Task Force (USMC)
MAHRS	Microflex Attitude & Heading Reference System
MAHRU	Microflex Attitude & Heading Reference Unit
MAI	Management Analysis Inc. (USA)
MAI	Ministry of Armament Industry (China)
MAID	Mobile Autonomous Intelligent Device
MAID/MILES	Magnetic Anti-Intrusion Detector / Magnetic Intrusion Line Sensor
MAIR	Maritime Air (NATO)
MAIREASTLANT	Maritime Air, Eastern Atlantic Command (NATO)
Maj	Major
Maj Gen	Major-General
MALE	Mando de Apoyo Logistico del Ejercito (Spain)
MALM	Master Air Loadmaster (Royal Air Force)
MALOS	Miniature Optical Laser Sight
MAMS	Mobile Air Movement Squadron (Royal Air Force)
MANCLOS	Manual Command to Line-Of-Sight
MANDATE	MCCIS Austere Northwood Database & Terminal Equipment
MANPADS	Manportable Air Defence System
MANS	Missile And Nudet Surveillance
MANTIS	Man-in-the-loop Target Interdiction System (UK)
MANTIS	Manpack Tactical Intelligence System
MAOT	Mobile Air Operations Team (UK Army)
MAOV	Mobile Artillery Observation Vehicle
MAP	Military Aid Program (USA)
MAP	Military Assistance Program (USA)
MAP	MILSTAR Advanced Processor (USA)
MAP	Multiple Aim Point (USAF)
MAP	Mutual Assistance Programme
MAPCO	Map Code
MAPS	Maritime Asset Planning System
MAPS	Mina Anti-pessoal de Plastico. Anti-personnel mine (Portugal)
MAPS	Mobile Aerial Port Squadron (USMAC)
MAPS	Modular Azimuth Positioning System
MAR	Minimally Attended Radar
MAR	Multiple / Multi-function Array Radar
MARA	Modular Architecture for Real-time Applications
MARAD	Maritime Administration
MARAIRMED	Maritime Air Forces, Mediterranean (NATO)
MARCLIP	Maritime Command Long Term Infrastructure Plan
MARCOM	Maritime Command (Canada)
MARCONFORLANT	Maritime Contingency Forces, Atlantic (NATO)
MARCS	MAC Airlift Reaction Communications System (USA)
MARDI	Mobile Advanced Robotic Defence Initiative
MARE	Miniature Analogue Recording Electronics
MAREC	Maritime Reconnaissance radar

MAREQ	Military Assistance Requirement
MARIN	Maritime Research Institute Netherlands
MARINTSUM	Maritime Intelligence Summary
MARMOSET	Marconi Mobile Satellite Earth Terminal
MARPAC	Maritime Forces, Pacific (Canada)
MARREP	Maritime Report
MARRES	Manual Radar-Reconnaissance Exploitation System
MARRS	Modular Armoured Repair & Recovery System
MARS	Magnetic Array Sensor System
MARS	Matra road to Air transportable Reconnaissance System (France)
MARS	Mid-Air Recovery System (US Navy)
MARS	Mid-Air Retrieval System (USAF)
MARS	Military Affiliate Radio System (USA)
MARS	Military Amphibious Reconnaissance System
MARS	Military Archive & Research Services (UK)
MARS	Minimally Attended Radar Station
MARS	Mobile Automatic Reporting Station
MARS	Modular Adaptable Radar Simulator
MARS	Modular Airborne Recording System
MARS	Multi-Access Retrieval System
MART	Mean Active Repair Time
MART	Mini-Avion de Reconnaissance Telepilote (France)
MART	Modular Air / Ground Remote Terminal
MARTEL	Missile Anti-Radar & Television
MARTHA	Maillage Anti-aerien des Radars Tactiques contre Helicopteres et Avions. Airspace management network (France)
MARV	Manoeuvring Re-entry Vehicle
MAS	Military Agency for Standardisation
MAS	Military Area Services
MASER	Microwave Amplification by Stimulated Emission of Radiation
MASH	Mobile Army Surgical Hospital (USA)
MASHRAG	Egypt, Jordan, Lebanon & Syria
MASINT	Measuring & Signature Intelligence
MASR	Multiple Antenna Surveillance Radar
MASRCLIP	Maritime Commander's Long-term Infrastructure Plan
MASS	Marine Air Support Squadron (USA)
MASS	Maritime Air Surveillance Sortie
MASS	Mission Avionics Sensor Synergism
MASST	Major Ship Satellite Terminal (USA)
MAST	Military Anti-Shock Trousers
MAST	Multi-Application Sonar Trainer
MAST	Multimission Airborne Surveillance Technology programme (USAF)
MASTACS	Manoeuvrability-Augmentation System for Tactical Air Combat Simulation (US Navy)
MASTER	Military Aircraft Satcoms Terminal
MASTER	Modular Acoustic Stimulator / Emulator
MASTT	MPS Action Speed Tactical Trainer
MASU	Multiple Acceleration Sensor Unit
MASWT	MPS Anti-Submarine Warfare Trainer

MAT ACCAT	Mobile Access Terminal Advanced Command & Control Architectural Test-bed (USA)
MatABw	Materialamt der Bundeswehr (Germany)
MATADOR	Multi-role Adaptive Tactical Auto-programmable Dynamic Radio System
MATCALS	Marine Air Traffic Control And Landing System (USA)
MATCH	Manned Anti-submarine Troop-Carrying Helicopter
MATCH	Medium Anti-Submarine Torpedo Carrying Helicopter
MATEL	Multiplexed Automatic Telephone Equipment
MATNET	Mobile Access Terminal Advanced Command & Control Architectural test-bed (USA)
MATO	Military Air-Traffic Operations (UK)
MATS	Marconi Acoustic Training System (UK)
MATS	Military Aircraft Target System
MATS	Mobile Automatic Telephone System
MATS	Model Aircraft Target System
MATS	Multiple Array Test Set
MATTS	Multiple Airborne Target Trajectory System
MATZ	Military Air Traffic Zone
MAU	Marine Amphibious Unit
MAV	Maintenance Assist Vehicle
MAVID	Modular Architecture for Video & Image Distribution system
MAW	Medium Anti-Tank Weapon
MAW	Military Airlift Wing (USAF)
MAW	Missile Approach Warning system
MAW	Mission Adaptive Wing
MAW	Mountain & Arctic Warfare
MAWTS	Marine Aviation Weapons & Tactics Squadron
max	Maximum
MBA	Main Battle Area
MBAR	Multiple Beam Acquisition Radar
MBAT	Multi-Beam Array Transmitter
MBB	Messerschmitt-Bolkow-Blohm (now part of Deutsche Aerospace) (Germany)
MBC	Military Budget Committee
MBC	Muzzle / Brake Compensator
MBFR	Mutual & Balanced Force Reductions
MBR	Marker Beacon Receiver
MBRLS	Multi-Barrel Rocket Launching System
MBRV	Manoeuvrable Ballistic Re-entry Vehicle
MBSS	Multi-Bank Staring System Sensor
MBT	Main Battle Tank
MC	Medium Capacity
MC	Medium Case bomb
MC	Military Committee (NATO)
MC	Mission Computer
MC	Mission-Capable
MC3	Multi-channel Crypto Controller
MCAIR	McDonnell Aircraft Co. (USA)
MCAS	Machinery Control And Surveillance

MCATT	Multi-weapon Combined Arms Tactical Trainer
MCC	Canada-US Military Co-operation Community, US Section
MCC	Management, Command & Control
MCC	Manual Control & Counter
MCC	Meteor Communications Corp. (USA)
MCC	Missile Control Console
MCC	Mobile Command Centre
MCCDC	Marine Corps Combat Development Command (USA)
MCCIS	Maritime Command, Control & Information System
MCCISWG	Military Command, Control & Information Systems Working Group
MCCP	Main Communications Control Panel
MCCS	Marconi Command & Control Systems
MCCS	Mine Countermeasures Control System
MCCU	Micro-Climate Cooling Unit
MCD	Marine Craft Detachment (Royal Air Force)
MCDEC	Marine Corps Development & Educational Command (USA)
MCDEC	Marine Corps School (USA)
MCDU	Multifunction Control & Display Unit
MCE	Micro Circuit Engineering Ltd (UK)
MCE	Modular Control Equipment (USA)
MCEB	Military Communications Electronics Board (USA)
MCEWG	Military Communications Electronics Working Group
MCF	Canon de Moyen Calibre Future (France)
MCFS	MPS Coastal Fortress Simulator
MCG	Master Control Group
MCG	Millimeter-wave Contrast Guidance
MCGS	Microwave Command-Guidance System
MCL	Mecanique Creusot-Loire (France)
MCLOS	Manual Command to Line-Of-Sight
MCM	Mine Countermeasures ship (USA)
MCMV	Mine Countermeasures Vessel
MCP	Military Construction Program (USAF)
MCP	Missile Control Panel
MCPE	Modular Collective Protection Equipment
MCR	Management Consulting & Research Inc. (USA)
MCR	Mobile Communication Radio
MCRC	Mobile Control & Reporting Centre
MCRV	Mechanised Combat Repair Vehicle
MCS	Machinery Control System
MCS	Maneuver Control System (USA)
MCS	Master Control Set
MCS	Master Control Station
MCS	Military Communications Systems
MCS	Missile Control System
MCS	Mobile Control Station
MCS	Multi-media Communications Station
MCSK	Mine Clearance System Kit (USA)
MCSR	Mission Completion Success Rate
MCT	Medium Combat Truck
MCT	MILAN Compact Turret (USA)

MCT	Military Combat Thrust
MCT	Moving Coil Transducer
MCTFIST	Marine Corps Tank Full Crew Interactive Simulator Trainer
MCTL	Military Critical Technologies List
MCTNS	Manportable Common Thermal Night Sight (USA)
MCTR	Message Center (USA)
MCU	Main Control Unit
MCU	Management Control Unit
MCU	Marine Craft Unit (Royal Air Force)
MCU	Master Control Unit
MCU	Missile Control Unit
MCU	Mission Control Unit
MCU	Modular Concept Unit
MCV	Mechanised Combat Vehicle
MCW	Modulated Continuous Wave
MCWS	Minor Caliber Weapon Station (USA)
MC&G	Mapping, Charting & Geodesy (USAF)
MC/FD	Modular Chaff / Flare Dispenser
MD	Methyldichloroarsine. Chemical warfare blister agent
MD	Military District
MDAP	Mutual Defense Assistance Program (USA)
MDC	Main Display Console
MDC	Maintenance Data Centre (Royal Air Force)
MDC	McDonnell Douglas Corp. (USA)
MDC	Mine Data Centre
MDC	Radio set control group (Spain)
MDE	Midland Diving Equipment Ltd (UK)
MDF	Maritime Defence Force
MDF	Mission Data File
MDF	Mission Degradation Factor (USAF)
MDF	Monopulse Direction Finding
MDHC	McDonnell Douglas Helicopter Co. (USA)
MDI	Material Dynamics Inc. (Canada)
MDI	Miss Distance Indicator
MDI	Modular Devices Inc. (USA)
MDIC	Malaysian Defence Industries Council
MDISI	McDonnell Douglas Information Systems International (UK)
MDJL	McDonnell Douglas Japan Ltd
MDK	Mashina Dorozhnoy Kopatelnoy. Trench digging machine (CIS)
MDLT	Mobile Data Link Terminal
MDMSC	McDonnell Douglas Missile Systems Co. (USA)
MDP	Maintenance Data Panel
MDP	Ministry of Defence Police
MDP	Modular Display Processor
MDP	Multi-Designation Protocol. Communications protocol
MDR	Medium Data Rate
MDR	Moyen de Deminage Rapide (France)
MDRI	Multi-purpose Display Repeater Indicator
MDS	Meteorological Data System
MDS	Modular Decontamination System

MDS	Modular Dispenser System
MDSL	Marconi Defence Systems Ltd (UK)
MDSSC	McDonnell Douglas Space Systems Co. (USA)
MDTM	Medium-Duty Tank Target Mechanism
MDTS	McDonnell Douglas Training Systems (USA)
MDTS	Megabit Digital Troposcatter Subsystem (USA)
MDTS	Mission Data Transfer System
MDU	Mine Distributing Unit
MDV	Mine Detection Vehicle
MDWP	Mutual Defense Weapons Program (USA)
MDZ	Missile Danger Zone
MD(G)T	Mission Data (Ground) Terminal
MD/WT	Marine Division / Wing Team (USA)
ME	Maintenance Error
ME	Manoeuvre Enhancement mode
ME	Mass Effectiveness
MEAS	Mechanical Engineering Aircraft Squadron (Royal Air Force)
MEB	Marine Expeditionary Brigade (USA)
MEC	Modular Electronics Concept
MECA	Missile Electronics & Computer Assembly
MECDL	Mission Equipment Control Data Link
Mech	Mechanical
MECO	Mechanical Equipment Co. Inc. (USA)
MECOM	Middle East Electronic Communications Exhibition
MECTS	Modular Electronic Combat Training System
MED	Message Exchange Device
MEDA	Military Emergency Diversion Airfield
MEDAX	Message Data Exchange terminal (USA)
MEDCENT	Allied Forces, Central Mediterranean (NATO)
MEDDS	Mechem Explosives & Drug Detection System (South Africa)
MEDE	Message Entry & Distribution Equipment
MEDEAST	Allied Forces, Southern Europe, Mediterranean East (NATO)
MEDEF	Middle East Defence & Security Exhibition
MEDEVAC	Medical Evacuation
MEDNOREAST	Allied Forces, Southern Europe, Mediterranean North East (NATO)
MEDOC	Allied Forces, Southern Europe, Mediterranean West (NATO)
MEDSOUTHEAST	Allied Forces, Southern Europe, Mediterranean South East (NATO)
MEECN	Minimum Essential Emergency Communications Network (USA)
MEF	Marine Expeditionary Force (USA)
MEF	Ministry of Economy & Finance (Uruguay)
MEFP	Ministere de l'Economie des Finances et du Budget (France)
MEFPMRS	Marine Expeditionary Force Primary Multi-channel Radio System (USA)
MEL	Magnesium Elektron Ltd (UK)
MEL	Missile Ejector Launcher
MELCO	Mitsubishi Electric Corp. (Japan)
MEMIC	Mobile Electro-Magnetic Incompatibility

MENS	Mission Element Need Statement (USA)
MEO	Military Electronics Office
MEOSS	Mobile Electro-Optical Surveillance System
MEOW	Multiple Engineering Order Wire
MEP	Medium External Pintle (Belgium)
MEP	Microelectronics Package
MEPU	Monofuel Emergency Power Unit
MER	Mission Evaluation Room
MER	Multiple Ejector Rack
MERA	Molecular-Electronics Radar
MERADCOM	Mobility Equipment Research & Development Command
MERE	Mortar Elevation & Ranging Equipment
MERLIN	Modular Ejection-Rated Low-profile Imaging for Night
MEROD	Message Entry & Read Out Device
MERZONES	Merchant shipping Zones
MESAR	Multi-function Electronically Scanned Adaptive Radar
MET	Marksmanship Expert Trainer
MET	Mission Event Timer
Met O	Meteorological Office (UK)
METIS	Meteorological Information System
METO	Maximum Except Take-Off
METOC	Meteorological / Oceanographic
MEU	Main Electronics Unit
MEU	Marine Expeditionary Unit (USA)
MEV	Medical Evacuation Vehicle (USA)
MEW	Microwave Early Warning
MEWPP	Modular Electronic Warfare Pre-Processor
MEWS	Microwave EW System
MEWS	Missile Early Warning Station
MEWS	Modular EW Simulator
MEWSG	Multinational / Maritime Electronic Warfare Support Group (NATO)
MEWSS	Mobile Electronic Warfare Support System
MEX	Microelectronics
MEXE	Military Engineering Experimental Establishment
MEZ	Missile Engagement Zone
MEZ	Mitteleuropaische Zeit. Central European Time
MF	Main Force
MF	Multi-Function
MF V	Materiel Division, Western Military Region Command (Sweden)
MFA	Movimento das Forcas Armadas. Armed Forces Movement (Portugal)
MFAR	Multi-Function Array Radar
MFBF	Multi-Function Bomb Fuze
MFC	Mortar Fire Controller
MFCC	Multi-Function Common Console
MFCD	Modular Flare / Chaff Dispenser
MFCD	MultiFunction Colour Display
MFCS	Missile Fire Control System
MFCS	Modified Fire Control System
MFD	Militarischer Frauendienst (Switzerland)

MFD(S)	Multi-Function Display (System)
MFF	Mixed Fighter Force
MFF	Munition Filling Factory
MFFC	Mixed Fighter Force Concept
MFHD	Multi-Function Head-down Display
MFMA	Multi-Function Microwave Aperture
MFMS	Military Flight Management System
MFN	Most Favoured Nation. IMF Clause
MFP	Medium Floor Pintle (Belgium)
MFR	Multi-Function Radar
MFRD	Moyen de Forage Rapide de Destruction. Mobile drilling machine (France)
MFS	Mottarraum-Feuerloschanlage. Engine compartment fire extinguishing system (Austria)
MFT	Metaalwaren-Fabriek Tilburg BV (Netherlands)
MG	Machine Gun
MG	Main Group (NATO)
MGARJS	Mobile Ground-to-Air Radar Jamming System
MGB	Medium Girder Bridge
MGCS	Missile Guidance & Control System
MGCS	Mobile Ground Control Station
MGF	Metallised Glass-Fibre chaff
MGGB	Modular Guided Glide Bomb
MGM	Master Group Multiplexer. Communications
MGM	Mobile surface attack Guided Missile
MgO	Magnesium oxide
MGO	Master General of the Ordnance
MGRS	Military Grid Reference System
MGS	Mobile Ground Station
MGT	Mobile Ground Terminals
MH	Minehunter
MHA	Mission Hydrographique L'Atlantique (France)
MHC	Materials Handling Crane
MHC	Minehunter, Coastal (USA)
MHI	Manual Hit Indicator
MHI	Mitsubishi Heavy Industries Ltd (Japan)
MHIDAS	Modular High Integration Distributed Architecture databus System
MHIP	Missile Homing Improvement Program (USA)
MHN	Moving Haven
MHQ	Maritime Headquarters
MHS	Message Handling System
MHS	Militarhogskolan (Sweden)
MHSV	Multi-purpose High Speed Vehicle
MHV	Miniature Homing Vehicle
MI	Marconi Instruments Ltd (UK)
MI	Maritime Interdiction
MI	Military Intelligence
MIA	Missing In Action
MIACAH	Mine Anti-Char d'Action Horizontale (France)
MIAMI	Microwave Ice Accretion Measurement Instrument

MIB	Mercantile Italo Britannica Srl (Italy)
MIC	Martinello Importing Co. Ltd (Canada)
MIC	Microwave Integrated Circuit
MIC	MLRS International Corp. (UK)
MICA	Missile d'Interception et de Combat Aerien. Combat & air intercept missile (France)
MICLIC	Mineclearing Line Charge
MICNS	Modular Integrated Communications & Navigation System (USA)
MICOM	Missile Command (US Army)
MICOS	Multifunctional Infra-red Coherent Optical Scanner
Micro-AIDS	Micro-Aircraft Integrated Data System
MICS	Manned Interactive Control Station
MICS	Mobile Integrated Communications System
MICV	Mechanised Infantry Combat Vehicle
Mid	Midshipman
MIDAS	Management Information, Data & Accounting System
MIDAS	Marine Inclination Differential Alignment System
MIDAS	Mine & Ice Detection / Avoidance System
MIDAS	Missile Defense Alarm System (USA)
MIDAS	Mobile Integrated Digital Automatic System
MIDL	Modular Interoperable Data Link
MIDS	Multifunction Information Distribution System (NATO)
MIDU	Missile Ignition Delay Unit
MIES	Multi-Imagery Exploitation System
MIFASS	Marine Integrated Fire & Air Support System
MIFV	Mechanised Infantry Fighting Vehicle
MiG	Mikoyan-Gurevich Aircraft (CIS)
MIG	MCE Interface Group
MIGITS	Miniature Integrated GPS/INS Tactical System
MIIRS	Modular Imagery Interpretation & Reporting System
MIL	Military
MILAN	Missile d'Infantrie Leger Anti-char. Light infantry anti-tank missile (France)
MILAS	Missile de Lutte Anti-Sous-marine. Anti-submarine missile
MILCAPSYS	Measuring Military Capability & Constraints Information System
MILCON	Military Construction
MILD	Magnetic-Intrusion Line Detector
MILES	Multiple Integrated Laser Engagement System (USA)
MIL-HDBK	Military Handbook
Milo M	Mellersta Militaromradet (Sweden)
Milo S	Sodra Militaromradet (Sweden)
Milo V	Vastra Militaromradet (Sweden)
MILREP	Military Representative (NATO)
MILSATCOM	Military Satellite Communications (USA)
MIL-SPEC	Military Specification
MILSTAMP	Military Standard Transportation And Movement Procedures
MILSTAN	Military Standard
MILSTAR	Military Strategic & Tactical Relay
MIL-STD	Military Standard
MILTRACS	Military Air Traffic Control System

MILU	Missile Interface & Logic Unit
MIM	Mobile-launched surface-to-air Missile
min	Minimum
MINDEF	Ministry of Defence (Singapore)
MINOS	Marconi Integrated Naval Operations System (UK)
MINT	Mutual Interference
MINTEX	M International Inc. (USA)
MIP	Missile Impact Predictor
MIPS	Medium Integrated Propulsion System (USA)
MIPS	Missile Impact Prediction Set
MIR	Multiple-target Instrumentation Radar
MIRA	MILAN Infra-Red Attachment
MIRA	Miniature Infra-Red Alarm
MIRACL	Mid Infra-Red Advanced Chemical Laser
MIRADCOM	Missile R&D Command (US Army)
MIRADOR	Minefield Reconnaissance & Detector System
MIRAGE	Microelectronic Indicator for Radar Ground Equipment
MIRC	Missile In-Range Computer
MIREL	Mortier a Injection Regeneratrice d'Ergol Liquide (France)
MIRTS	Modular Infra-Red Transmitting System
MIRV	Multiple Independently Targetable Re-entry Vehicle
MIS	Management Information System
MIS	Message Input Segment
MIS	Military Information System
MISREP	Mission Report
MISS	Missile Intercept Scoring System
MIST	Modular Interoperable Surface Terminal
MIST	Mosaic Infra-red Sensor Technology
MISTIGRI	Mobile Integrated Surveillance of Tactical Information Gathered by Remote Informants (France)
MITAS	Multi-sensor Imaging Technology for Airborne Surveillance
MITI	Ministry of International Trade & Industry (Japan)
MITS	Mobile Independent Target System
MIU	Missile Interface Unit
MIU	Multiplex Interface Unit
MiWS	Minenwerfersystem (Germany)
MIZ	Materialinformationszentrum Gesellschaft fur Logistik mbH (Germany)
MK	Morrison Knudsen Corp. (USA)
MKC	Multiple Kill Capacity
MKE	Makina ve Kimya Endustrisi Kurumu (Turkey)
MKS	Multi-Kommunikations-System (Germany)
MKT	Maschinen Kanonen Trainer (Germany)
MKV	Miniature Kill Vehicle
ML	Minelayer
ML	Missile Launcher
MLB	Main Lobe Blanking
MLBM	Modern Large Ballistic Missile
MLC	Main Lobe Cluster
MLC	Military Load Classification

MLC	Modular Load Carrier
MLE	Missile Launch Envelope
MLF	Multilaminar Film
MLF	Multi-Lateral Fleet
MLF	Multi-Lateral nuclear Force
MLFS	Modular Lightweight FLIR System
MLI	Mid-Life Improvement
MLLV	Medium-Lift Launch Vehicle
MLM	Modern Labelling Methods Ltd (UK)
MLMS	Mulit-purpose Lightweight Missile System
MLPP	Multi-Level Precedence & Pre-emption
MLPRF	Modular Low Power Radio Frequency
MLPS	Multilink Processing System
MLRS	Multiple-Launch Rocket System (USA)
MLS	Microwave Landing System
MLS	Microwave Limb Sounder
MLS	Multi-Level Security
MLT	Missile Loader Transport
MLT	Munitions Lift Trailer
MLU	Mid-Life Update
MLU	Monitor & Logic Unit
MLV	Memory Loader & Verifier
MLV	Mobile Launch Vehicle
MLVW	Medium Logistic Vehicle, Wheeled
MLW	Maximum Landing Weight
MM	Minuteman series ballistic missile (USA)
MMA	Multimission Aircraft
MMBF	Mean Miles Between Failures
MMC	Materiel Management Center (US Army)
MMC	Mission Management Computer
MMD	Multimode Display
MMDS	Moving Map Display Equipment
MMDS	Multi-Mode Seeker Demonstration project
MMFF	Multimode Fire & Forget
MMH	Monomethyl Hydrazine
MMI	Man Machine Interface
MMIC	Monolithic Microwave Integrated Circuit
MMIMS	Modular Multi-Influence Minesweeping System
MMLC	Medium Mobility Load Carrier
Mmo	Maximum operating Mach number
MMP	Maintenance Monitoring Panel
MMP	Modular Mission Payload
MMR	Minimum Military Requirement
MMRPV	Multimission RPV
MMS	Mast-Mounted Sight
MMS	Missile Management System
MMS	Mycon Marketing Services Ltd (UK)
MMS1	Mobile Mast Spectrometer 1
MMSS	MPS Multibeam Sonar Simulator
MMT	Miniature Mobile Target

MMU	Memory Management Unit
MMV	Modular Military Vehicle
MMV	Multi-Mission Vehicle
MMW	Medium Multi-purpose Wheeled
MMW	Millimetre Wave
MMWR	Millimetric Wave Radar
MNC	Major NATO Command
MND	Ministry of National Defence
MND	Mission Need Documents (NATO)
MND-C	Multi-National Division, Centre. To be formed between October 1993 & April 1994 as part of NATO's new ARRC
MND-S	Multi-National Division, South. Due to become operational in 1995, forming part of NATO's new ARRC
MNF	Maritime Nuclear Forces
MNFP	Multinational Fighter Program (USA)
MNK	Marine Nationale Khmer (Cambodia)
MNLF	Moro National Liberation Front
MNPS	Minimum Navigation Performance Specification
MNR	Mozambique National Resistance
MNS	Mine Neutralisation System
MNS	Mission Needs Statement
MOA	Memorandum Of Agreement
MOA	Military Operations Area
MOAT	Minewarfare Operational Analysis Tool
MOB	Main Operational Base (USAF)
MOBA	Mobility Operations For Built-up Areas (US Army)
MOBSS	Mobility Support Squadron
MOC	Modular Operations Centre
MOC	Multifunction Operator Console
MoD	Ministry of Defence (UK)
MOD CIS	MoD Communication Information System
MOD DIG	Modular Digital Image Generator
MODA	Ministry of Defense & Aviation (Saudi Arabia)
MODAS	Modular Data Acquisition System
MODELS	Modernisation of the Defence Logistics Systems
MODEM	Modulator / de-modulator
MODIR	Modulated Infra-Red jammer
MODUK	Ministry of Defence UK (NATO)
MOD(PTS)	Ministry Of Defence Police Training School (UK)
MOE	Measure Of Effectiveness
MOFA	Multi-Option Fuze Artillery
MOGUL	Modular Gun Laying System
MOLAR	Mortar / Artillery Locating Radar
MOLF	Modular Laser Fire Control
MOM	Mission Oceanographique de la Mediterranee (France)
MOM	Multirole OTO Munition
MOMCOMS	Man-On-the-Move Communications System (USA)
MOMS	Modular Opto-electronic Multi-spectral Scanner
MON	Minnoye Oskolochonym Napravleniem. Anti-personnel mine with directional fragmentation (CIS)

MONAB	Mobile Naval Airbase, on shore (US Navy)
MONOHUD	Monocular Head-Up Display
MOPMS	Modular Pack Mine System
MOPS	Minimum Operating Performance Standard
MOPTAR	Multiple-Object Phase Tracking And Ranging
MOR	Military Operational Requirement
MORS	Minefield & Ordnance Recovery Management System
MOS	Meridian Ocean Systems (USA)
MOS	Military Occupational Specialty qualifications (USA)
MOS	Monitor Only crew Station
MOSFET	Metal Oxide Silicon Field Effect Transistor
MOSP	Multi-mission Optronic Stabilised Payload (UAV)
MOST	Mission-Orientated Simulator Training (USA)
MoT	Ministry of Transport
MOTEC	Maritime Operations Tactical Evaluation Centre
MOTIS	Message Orientated Text Interchange System
MOT-key	Modified One-Time key
MOTR	Multiple Object Tracking Radar
MoU	Memorandum of Understanding
MOUT	Mobile Operation in Urban Terrain
MOVTAS	Modified Visual Target Acquisition System
MOWAM	Mobile Water Mine
MP	Manoeuvre Programmer
MP	Manpower & Personnel (USAF)
MP	Metal-Piercing
MP	Military Police
MP	Mornaricka Pjesadija. Croatian naval infantry
MP	Multipurpose
MPA	Man-Powered Aircraft
MPA	Maritime Patrol Aircraft
MPAA	Multifunctional / Multi-beam Phased-Array Antenna
MPAT	Multi-Purpose Anti-Tank
MPBA	Multiple Practice Bomb Adapter
MPC	Military Personnel Centre
MPC	Missile Practice Camp
MPC	Multi-Purpose Console
MPCD	Multi-Purpose Colour Display
MPCS	Mission Planning & Control Station
MPD	Maximum Permitted Dose
MPD	Multipurpose Display
MPDS	Message Processing & Distribution System (USA)
MPDS	Missile-Piercing, Discarding Sabot
MPDS	Multi-Purpose Decontamination System
MPE	Midi Provence Electronique (France)
MPFF	Multi-Purpose Fighter Facility
MPGS	Mobile Protected Gun System (USA)
MPH	Medium Pintle Head (Belgium)
MPHT	Missile Potential Hazard Team
MPI	Multiprotocol Interface
MPIM	Multi-Purpose Individual Munition

MPLA	Popular Movement for the Liberation of Angola
MPM	Multi-Purpose Missile (USAF)
MPMS	Missile Performance Monitoring System
MPN	MSE Packet Network
MPO	Mission Planning Officer (Royal Air Force)
MPOC	Multipurpose Operators Console
MPP	Massively Parallel Processor
MPPU	Multi-Purpose Processor Unit
MPR	Medium Power Radar
MPR	Military Power Reserve
MPS	Main Propulsion System
MPS	Maritime Prepositioning Ships (USA)
MPS	Micro Processor Systems A/S (Norway)
MPS	Mission Payload Subsystem
MPS	Mixed-Propellant System
MPS	Motor Pump Set
MPS	Multiple Protective Shelter
MPT	Multi-Purpose Tracer
MP-T-SD	Multi-Purpose Tracer, Self-Destroying
MPU	Missile Power Unit
MPU	Mission Programming Unit
MPWS	Mobile Protected Weapon System (USA)
MP(K)	Machine Pistole (Krimmi) (Germany)
MQLF	Mobile Quick-Look Facility
MR	Maritime Reconnaissance
MR	Marker Ranger
MR	Medium Range
MR	Military region
MR	Simulated Mustard. Chemical warfare blister agent
MRAAM	Medium-Range Air-to-Air Missile
MRAD	Multiple Range Alignment Device
MRASM	Medium-Range Air-to-Surface Missile
MRB	Medium Range Blast
MRBF	Mean Rounds Before Failure
MRBM	Medium-Range Ballistic Missile
MRC	Message-based Reliable Channel
MRC	Microwave Radio Corp. (USA)
MRCA	Multi-Role Combat Aircraft
MRCLOS	Missile Reference Command-to-Line-Of-Sight
MRCS	Mobile Reporting & Control System
MRCS	Multiple RPV Control System
MRD	Motor-Rifle Division
MRDFS	Man-transportable Radio Direction Finding System
MREL	Mining Resource Engineering Ltd (Canada)
MRF	Multi-Role Fighter (USA)
MRF	Multi-Role Fuze
MRL	Maritime Rear Link
MRL	Modular Rocket Launcher
MRL	Multiple Rocket Launcher

MROS	Multirole Operations Cabin
MRPV	Mini Remotely Piloted Vehicle
MRR	Medium Range Recovery
MRR	Missile Reliability Restoration
MRR	Multi-Role Radar
MRS	Mobile Reception System
MRS	Multiple Rocket System
MRS	Multirole System
MRS	Muzzle Reference System
MRSAM	Medium-Range Surface-to-Air Missile
MRSR	Magazine Ready Service Rings
MRSR	Multiple Role Survivable Radar
MRT	Manpack Radio Telephone
MRT	Miniature Receiver Terminal
MRT	Mobile Radio Terminal
MRT	Mobile Repair Team
MRT	Multi-Radar Tracking
MRT	Multirole Radio Telephone
MRTAF	Marshall of the Royal Air Force
MRTFL	Medium Rough Terrain Forklift
MRTI	Multi-Role Thermal Imager
MRT-S	Multi-Role Turret System
MRTU	Multiple Remote Terminal Unit
MRU	Maintenance Recorder Unit
MRV	Maintenance-Recovery Vehicle
MRV	Multiple Re-entry Vehicle
MRVC	Multiple Rate Voice Card
MRVR	Mechanised Repair & Recovery Vehicle (USA)
MRW	Maximum Ramp Weight
MS	Maritime Surveillance
MS	Measurement Systems Inc. (USA)
MS	Methyl Salicylate
MS	Military Secretary, Department of (UK)
MS	Military Standards (USA)
MSA	Maritime Safety Agency (Japan)
MSA	Mobile Subscriber Access
MSA	Mutual Security Agency
M-SAM	Medium Surface-to-Air Missile
MSAM	Medium-range Surface-to-Air Missile
MSAR	Miniature Synthetic Aperture Radar
MSAT-Air	Multi-Sensor Aided Targeting – Airborne
MSAW	Minimum Safe Altitude Warning
MSB	Minesweeping Boats (USA)
MSBS	Mer-Sol Balistique Strategique. Submarine-launched ballistic missile (France)
MSC	Major Subordinate Commander (NATO)
MSC	Military Sealift Command (USA)
MSC	Minesweeper, Coastal
MSCE	Modular, Standard Control Electronics for OFSA
MSCS	Multiservice Communications System project (USA)

MSD	Minesweeping Drone (USA)
MSD	Multisensor Display
MSDS	Marconi Space & Defence System (now Marconi Radar & Defence Systems) (UK)
MSE	Malaysia Shipyard & Engineering Sdn Bhd
MSE	Mobile Subscriber Equipment (USA)
MSER	Multiple Stores Ejector Rack
MSF	Militarily Significant Fallout
MSF	Multisensor Fusion
MSFC	Marshall Space Flight Center (USA)
MSFP	Mosaic-Staring Focal Plane
MSG	Message
MSG	Minensuchgerat (Germany)
MSGL	Multi-Salvo Grenade Launcher
MSI	Marine Specialty Inc. (USA)
MSI	Medium-Scale Integration
MSI	Minesweeper, Inshore (USA)
MSIP	Multinational Staged Improvement Programme
MSIS	Multi-sensor Stabilised Integrated Sensor
MSK	Minimum Shift Keying
MSM	Minesweeper, River (USA)
MSMS	Magnetic Signature Measurement System
MSM/W	Minenstreumittel-Werfer (Germany)
MSO	Master Specification Officer
MSO	Minesweeper, Ocean
MSOW	Modular Stand-Off Weapons
MSPT	Marine Silent Power Transmission
MSR	Minesweeper, Patrol (USA)
MSR	Missile Site Radar
MSR	Mobile Sea Range
MSR	Modular Survivable Radar
MSR	Multi-beam Multi-mode Surveillance Radar
MSR	Multisensor Reconnaissance
MSRS	Miniature Sonobuoy Receiver System
MSRT	Mobile Subscriber Radiotelephone Terminals
MSS	Mine Search System
MSS	Miniature Surveillance System
MSS	Missile Support Subsystem
MSS	Mission Simulator System
MSS	Mission Support System
MSS	Mobile Satellite System
MSS	Mobile Subscriber Equipment System
MSSC	Maritime Surface Surveillance Capability
MSSC	Medium Seal Support Craft (USA)
MSSR	Monopulse Secondary Surveillance Radar
MST	Microprocessor Simulation Technology
MST	Missile Surveillance Technology
MST	Mobile Systems Technology (Belgium)
MST	Mountain Standard Time
MST	Multi-Sensor Tracking

MSTAR	Manportable Surveillance & Target Acquisition Radar
MSTART	Missile System to Attack Relocatable Targets
MSU	Mass Storage Unit
MSU	Mode Selector Unit
MSV	Micro Surveillance system
MSV	Modular Support Vehicle
MSWG	Military Strategy Working Group (NATO)
MSX	Midcourse Space Experiment
MT	Mechanical Time
MT	Mechanical Transport
MT	Military Transport
MT	Missile Target
MT	Multi-frequency Transducer
MTA	Mackenzie Tribbeck Associates Ltd (UK)
MTA	Military Training Area
MTACS	Marine Tactical Air Control System
MTAD	Multi-Trace Analysis Display
MTAE	Multiple-Time-Around Echoes
MTAS	Millimetric Target Acquisition System
MTAS	Modular Target Acquisition System
MTAS	Multi-sensor Target Acquisition System
MTAT	Maktab Turus Angkatan Tentera (Malaysia)
MTB	Missile Torpedo Boat
MTB	Mobility Test-Bed
MTB	Motor Torpedo Boat
MTBCF	Mean Time Between Critical Failures
MTBF	Mean Time Between Failures
MTBR	Mean Time Between Repairs
MTBSF	Mean Time Between Significant Failures
MTCAS	Marine Corps Tactical Command And Control System
MTCC	Modular Tactical Communication Centre
MTCR	Missile Technology Control Regime
MTD	Moving Target Detection
MTDS	Marine Tactical Data System (USA)
MTI	Moving Target Indicator
MTI	Multiple Target Interception
MTIA	Metal Trades Industry Association of Australia
MTIAC	Manufacturing Technology Information Analysis Center (USA)
MTL	Magnetic Tape Loader
MTL	Materials Technology Laboratory (USA)
MTLS	Magazine Torpedo Launch System
MTLV	Missile Transport & Loading Vehicle
MTM	Moving Target Mechanism
MTMA	Military Terminal Manoeuvring Area
MTMC	Military Traffic Management Command
MTMT	Multiple Target & Missile Tracker
MTM(WD)	Moving Target Mechanism (Winch Driven)
MTO	Military Transport Officer (UK)
MTOGW	Maximum Take-Off Gross Weight
MTR	Marked Target Receiver

MTR	Military Training Route
MTR	Missile Tracking Radar
MTR	Mobile Test Rig (USA)
MTR	MTU Turbomeca Rolls-Royce GmbH (Germany)
MTRACS	Multiple input Tracking Control System
MTRE	Missile Test & Readiness Equipment
MTS	Maritime Tactical School
MTS	Marked Target Seeker
MTS	Military Tactical Systems
MTS	Modular Tactical Switch
MTS	Moving Target Simulator
MTSQ	Mechanical Time & Superquick
MTSS	Mine Warfare Tactical Support System
MTT	Medium Tactical Truck
MTT	Moving Turning Target
MTT	Multiple Target Tracking
MTTF	Microbuoy Transportable Test Facility
MTTR	Mean Time-To-Repair
MTTR	Multi-Target Tracking Radar
MTTS	Multiplexed Tactical Telephone System
MTTS	Multi-Task Training System
MTU	Magnetic Tape Unit
MTU	Mostoykladchik Tankoviy Ustroystvo. Armoured Bridgelayer (CIS)
MTU	Motoren-und-Turbinen-Union (Germany)
MTV	Medium Tactical Vehicle
MTVT	Medium Tactical Vehicle Trailer
MUDAS	Modular Universal Data Acquisition System
MuFAR	Multi-target Field Artillery Radar (India)
MUFTI	Minimum Use of Force Tactical Intervention
MULE	Modular Universal Laser Equipment (USA)
MULTOTS	Multiple Unit Link-11 Test & Operational Training System (USA)
MULTS	Mobile Universal Link 11 Translator System
MUMS	Modular Underwater Measurement System
MURFAAMCE	Mutual Reduction of Forces & Armaments & Associated Measures in Central Europe
MUSL	Marconi Underwater Systems Ltd (UK)
MUTES	Multiple Threat Emitter System
MUU	Main User Unit
Mux	Multiplexer
MV	Muzzle Velocity
MVEE	Military Vehicles & Engineering Establishment
MVLD	Man Worn Laser Detector
MVLT	Morse / Voice Language Trainer
MVP	Military VME Processor
MVP	Motor Vehicle Plant
MVPS	Multiple Vertical Protective Shelter
MVR	Muzzle Velocity Radar
MW	Medium Wave
MW	Microwave
MW	Mine Warfare

MWA	Multiple Weapons Adapter
MWB	Motorenwerk Bremerhaven GmbH (Germany)
MWDO	Minewarfare Tactical Development Group (UK)
MWDP	Master Warning Display Panel
MWDP	Mutual Weapons Development Program (USA)
MWDS	Man-Worn Detector System
MWHQ	Mobile War Headquarters
MWR	Milimeter-Wave Radar
MWR	Mine Warfare Range
MWS	Manned Weapon Station (USA)
MWS	Master Warning System
MWS	Modular Workstation
MWT	Medium Wheeled Tractor
MWVP	Minor War Vessels Programme (UK)
MWVS	Mission Weapon Visionics System
MX	Missile Experiment or Missile-X (advanced ICBM)
MZFW	Maximum Zero-Fuel Weight
M&ES	Magnetic & Electromagnetic Silencing
M/S CPO	Master / Senior Chief Petty Officer (US Navy)

N

NA	Numerical Aperture
NAA	North Atlantic Assembly
NAADM	North American Air Defense Modernization
NAAFI	Navy, Army & Air Force Institutes (UK)
NAAG	NATO Army Armaments Group
NAAWS	NATO Anti-Air Warfare System
NABS	NATO Air Base Satcom
NABSE	Norwegian Artillery Battery Survey Equipment
NAC	NATO Alert Committee
NAC	Naval Air Command (UK)
NAC	Naval Avionics Center (USA)
NAC	Network Access Controller
NAC	North Atlantic Council (NATO)
NACA	National Advisory Committee for Aeronautics (USA)
NACC	North Atlantic Cooperation Council
NACCB	National Accreditation Council for Certification Bodies
NACISA	NATO Communications & Information Systems Agency (formerly NICSMA)
NACISC	NATO Communications & Information Systems Committee
NACISO	NATO Communications & Information Systems Organisation
NACMA	NATO ACCS Management Agency
NACMO	NATO ACCS Management Organisation
NADC	NATO Air Defence Committee
NADC	Naval Air Development Center (USA)
NADEEC	NATO Air Defence Electronic Environment Committee
NADEFCOL	NATO Defence College
NADEP	Naval Aviation Depot (USA)
NADGE	NATO Air Defence Ground Environment
NADL	Navy Avionics Development Laboratory (USA)
NADREPS	National Armaments Directors Representatives (NATO)
NADS	National Armaments Directors (NATO)
NAEDS	Non-Aqueous Equipment Decontamination System
NAEGIS	NATO Airborne Early-warning Ground Environment Integration System
NAEW	NATO Airborne Early Warning
NAEWF	NATO Airborne Early Warning Force
NAF	Naval Air Facility (USA)
NAFAG	NATO Air Force Armaments Group
NAG	Netherlands Aerospace Group
NAHEMA	NATO Helicopter Management Agency
NAIAD	Nerve Agent Immobilised enzyme Alarm & Detector (UK)
NAIC	NATO Intelligence Centre
NAIR	Naval Air Systems Command headquarters (USA)
NALLADS	Norwegian Army Low Level Air Defence System
NAMC	Nanchang Aircraft Manufacturing Co. (China)
NAMFI	NATO Missile Firing Installation
NAMMA	NATO MRCA Development & Production Management Agency
NAMSA	NATO Maintenance & Supply Agency

NAMSO	NATO Maintenance & Supply Organisation
NAPATMO	NATO Patriot Management Office
NAPC	Naval Air Projects Co-ordination office (USA)
NAPMA	NATO Airborne Early Warning & Control Programme Management Agency
NAPMO	NATO Airborne Early Warning & Control Programme Management Organisation
NAPO	NATO Production
NAPR	NATO Armaments Planning Review
NAPS	Nerve Agent Pre-treatment Tablets
NARF	Naval Air Reserve Force (USA)
NAS	National Aerospace Standards (USA)
NAS	Naval Air Station (USA)
NAS	Newport Aeronautical Sales (USA)
NASA	National Aeronautics & Space Administration (USA)
NASAMS	Norwegian Advanced Surface-to-Air Missile System
NASC	Naval Air Systems Command (USA)
NASCOM	NASA Communications network (USA)
NASDA	National Space Development Agency (Japan)
NASF	Navigation & Attack Systems Flight (Royal Air Force)
NASP	National Aero-Space Plane (USA)
NASP	Navy Airship Program (USA)
NASPO	Naval Air System Program Office (USA)
NASR	Naval & Air Staff Requirement
NASS	Naval Anti-Submarine School
NASSCO	National Steel & Shipbuilding Co. (USA)
NAST	Naval & Air Staff Target
NASWDU	Naval Air / Sea Warfare Development Unit
NAT	Northern Airborne Technology Ltd (Canada)
NATC	Naval Air Training Command (USA)
NATC	Nevada Automotive Test Center (USA)
NATINADS	NATO Integrated Air Defence System
NATIP	NATO Information & Press office
NATO	North Atlantic Treaty Organization
NAU	Network Access Unit
NAUTEL	Nautical Electronic Laboratories Ltd (Canada)
NAUTIC	Naval Autonomous Intelligent Console
NAUTIS	Naval Autonomous Information System (UK)
nav	Navigation
NAVAID	Navigation Aid
NAVAIRSYSCOM	Naval Air Systems Command (USA)
NAVAR	Combined Navigation And Radar system
NAVBALTAP	Allied Naval Forces, Baltic Approaches (NATO)
NAVCOMPARS	Naval Communications Processing And Routeing System (USA)
NAVDAC	Naval Data Automation Command (USA)
NAVEX	Navigation Exercise
NAVFACENGCOM	Naval Facilities Engineering Command (USA)
NAVHARS	Navigation, Heading & Attitude Reference System
NAVICERT	Navy Certificate
NAVMACS	Naval Modular Automated Communications System (USA)

NAVMATINST	Naval Materiel Command Instruction
NAVMIC	Naval Maritime Intelligence Center (USA)
NAVMOVE	Navigational Movements
NAVNON	Allied Naval Forces, North Norway (NATO)
NAVOBSY	Naval Observatory (USA)
NAVOCFORMED	Naval On-Call-Force, Mediterranean (NATO)
NAVSEASYSCOM	Naval Sea Systems Command (USA)
NAVSITSUM	Naval (own) Situation Summary
NAVSONOR	Allied Naval Forces, South Norway (NATO)
NAVSOUTH	Allied Naval Forces, Southern Europe (NATO)
NAVSPASUR	Naval Space Surveillance System (USA)
NAVSSES	Naval Ship System Engineering Station (USA)
NAVSTAR	Navigation Satellite Timing & Ranging (see GPS)
NAVSTARCODE	Naval Staff Target And Requirement Code
NAVSUPSYSCOM	Naval Supply Systems Command (USA)
NAVTAG	Naval Tactical Game
NavWASS	Navigation & Weapon Aiming Subsystem
NAW	Night / Adverse Weather
NAWACS	NATO AWACS
NAWAS	National Warning System (USA)
NAWC	Naval Air Warfare Center (USA)
NAWST	Night Attack Weapon Systems Trainer
NB	Naval Base/s
NBC	Nuclear, Biological & Chemical (warfare)
NBCD	Nuclear, Biological & Chemical Defence
NBCF	Nuclear, Biological, Chemical & Fire
NBCPC	Nuclear, Biological & Chemical Protective Cover
NBCRS	Nuclear, Biological & Chemical Reconnaissance System
NBCSS	Nuclear, Biological & Chemical Shelter System
NBMR	NATO Basic Military Requirement
NBS	Narrow Beam Sounder
NBSS	Narrowband Switching System
NBSV	Narrowband Secure Voice
NC	Narrow Coverage antenna
NC	Nitrocellulose
NC	Node Centre
NCA	National Command Authority (USA)
NCA	National Control Authority
NCAP	Night Combat Air Patrol
NCAPC	NATO Conventional Armaments Planning Committee
NCARC	NATO Conventional Armament Review Committee
NCB	NATO Codification Bureau
NCC	Network Control Centre
NCC	Northwood Control Centre
NCCIS	NATO Command, Control & Information System
NCDU	Navigational Control & Display Unit
NCE	Navigation & Command Equipment
NCEB	NATO Communications & Electronics Board
NCF	NATO Composite Force
NCI	Navigational Control Indicator

NCISC	Naval Counter-Intelligence Support Center (USA)
NCO	Non-Commissioned Officer
NCP	Network Control Points
NCS	NATO Codification System
NCS	Naval Combat System
NCS	Naval Communications System (USA)
NCS	Naval Control of Shipping (UK)
NCS	Network Control Station
NCS	Network Control Subsystem
NCS	Node Centre Switch
NCT	NATO Comparative Test programme
NCT	Network Control Terminal
NCT	Non-Co-operative Target
NCTR	Non-Co-operative Target Recognition
NCTS	Naval Computer & Telecommunications Station (USA)
NCU	Naval Communications Unit, Washington (USA)
NCU	Navigation Computer Unit
NCWA	NATO Wartime Civil Agencies
NC/AC-EC	Narrow Coverage / Area Coverage to Earth Coverage
NC/AC-NC/AC	Narrow Coverage / Area Coverage to Narrow Coverage / Area Coverage
ND	Navigation Display
ND	Navy Department
ND	Non-Detectable
Nd YAG	Neodymium-doped Yttrium Aluminium Garnet
NDA	National Defense Area (USA)
NDA	Naval Discipline Act (UK)
NDAC	Nuclear Defence Affairs Committee
NDB	Non Directional Beacon
NDB	Nuclear Depth Bomb
NDDN	Norwegian Defence Digital Network
NDE	Non-Destructive Evaluation
NDEW	Nuclear Directed Energy Weapon
NDEWT	Nuclear Directed Energy Weapon Technology programme (USA)
NDI	Non-Developmental Item
NDIC	National Defence Industries Council
NDN	Naval Digital Network
NDPO	National Defence Programme Outline (Japan)
NDRF	National Defence Reserve Fleet
NDS	Navigation Data Systems Inc. (USA)
NDS	Nuclear Detection / Detonation System
NDSTC	National Defence Science & Technology Commission (China)
NDSTIC	National Defence Science Technology & Industry Commission (China)
NDT	Non-Destructive Testing
NDU	National Defense University (USA)
NDU	Navigation Display Unit
NDV	Nuclear Delivery Vehicle
NEACP	National Emergency Airborne Command Post (USA)

NEARTIP	Near-Term Improvement Programme
NEB	Nuclear Exoatmospheric Burst
NEC	Northern European Command (NATO)
NEDD	NATO & European Defence Directorate
NEDPS	Nacken Electronic Data Processing System (Sweden)
NEFC	NATO Electronic Warfare Fusion Cell
NEFMA	NATO EFA Management Agency
NEFMO	NATO EFA Development Production & Logistics Management Organisation
NEI	Northern Engineering Industries plc (UK)
N-EMP	Nuclear Electromagnetic Pulse
NERVA	Nuclear Energy for Rocket Vehicle Applications
NES	Naval Engineering Standard
NES	Night Effects Simulator
NESC	Naval Electronics Systems Command (USA)
NESP	Navy EHF Satcom Program (USA)
NESS	Naval Engineering Support System
NESSEC	Naval Electronics Systems Security Engineering office (USA)
NET	Normal Environmental Temperature
NETC	Naval Education & Training Center (USA)
NETPMSA	Naval Education & Training Program Management Support Activity (USA)
NETSYO	Network Security Officer
NETTO	Network Training Officer (Royal Navy)
NEWAC	NATO Electronic Warfare Advisory Committee
NEWTS	Naval Electronic Warfare Training System
NEXRAD	Next-generation Radar
NFCS	Nuclear Forces Communications Satellite
NFH	NATO Frigate Helicopter
NFIP	National Foreign Intelligence Programme
NFM	North Finding Module
NFO	Naval Flight Officer (USA)
NFOV	Narrow Field-Of-Vision
NFR	NATO Frigate Replacement
NFT	Norsk Forsvarsteknologi A/S (Norway)
NFT UK	Norsk Forsvarsteknologi A/S (UK)
NFTS	Naval Fixed Telecommunications System
NG	Nouvelle Generation. New generation (France)
NGASR	Navy, Army & Air Staff Requirement
NGAST	Navy, Army & Air Staff Target
NGB	National Guard Bureau (USA)
NGL	Normalair-Garrett Ltd (UK)
NGPS	Navstar Global Positioning System
NGR	Night-Goggle Readable
NGS	Naval Gun Support
NGSFO	Naval Gunfire Support Forward Observer (UK)
NGV	Nozzle Guide Vane
NGW	Nuclear Gravity Weapon
NH-90	NATO Helicopter for the 1990's
NHI	NATO Helicopter Industries

NHMO	NATO HAWK Management Office
NHPLO	NATO HAWK Production & Logistics Organisation
NHR	Naval High Refresh display system
NH_4 ClO_4	Ammonium Perchlorate
NI	Naval Infantry
NIAG	NATO Industrial Advisory Group
NIC	Naval Intelligence Command (USA)
NiCd	Nickel Cadmium
NICS	NATO Integrated Communications System
NICS COA	NICS Control Operating Authority
NICSMA	NATO Integrated Communications System Management Agency (now NACISA)
NICSO	NATO Integrated Communications System Organisation
NIDIR	Nike Digital Instrumentation Radar
NIDS	National Institute for Defense Studies (Japan)
NIF	National Islamic Front (Sudan)
NII	Nuclear Installations Inspectorate
NIID	Netherlands Defense Manufacturers Association
NILU	National Intelligence Liaison Unit
NIMCSSC	National Military Command Systems Support Center (USA)
NIMIC	NATO Insensitive Munitions Information Centre
NIMP	NATO Interoperability Management Plan
NIPPI	Japan Aircraft Manufacturing Co. Ltd
NIPSSA	Naval Intelligence Processing Systems Support Activity office (USA)
NIR	Near Infra-Red
NIREX	Nuclear Industry Radioactive Waste Executive
NIS	NATO Identification System
NIS	NATO International Staff
NISSPO	NATO Identification System Special Project Office
NITE-OP	Night Imaging Through Electro-Optics
NITF	National Imagery Transmission Format (USA)
NITO/W	National Intelligence Tasking Office, Warning & crisis management (USA)
NIU	Network Interface Unit
NJ	Norsk Jetmotor A/S (Norway)
NK	N. Kioleides SA (Greece)
NLCM	Non-Lethal Countermeasures
NLIF	Non-Linear Interference Filter
NLOS	Non Line-Of-Sight
NLOS-CA	Non-Line-Of-Sight – Combined Arms
NLR	Nationaal Lucht-en Ruimtevaartlaboatorium (Netherlands)
NLS	New Launch System (NASA/US Air Force)
NLUS	Navy League of the United States
NMC	Naval Materiel Command (USA)
NMC	Naval Missile Center (USA)
NMC	Not Mission-Capable
NMCC	National Military Command Center (USA)
NMCRC	Navy Marine Corp Reserve Center (USA)
NMCS	National Military Command System (USA)
NMI	Nuclear Metals Inc. (USA)

NMIC	National Military Intelligence Center (USA)
NML	Naval Multiple Launcher
NMMS	Navy Mast Mounted Sight (USA)
NMR	National Military Representative to SHAPE (NATO)
NMR	Nuclear Magnetic Resonance
NMS	Network Management System
NMU	Navigation Management Unit
NMWS	Naval Meroka Weapon System (Spain)
NNAG	NATO Naval Armaments Group
NNIG	Netherlands Naval Industries Group
NNK	Non-Nuclear Kill
NNMSB	Non-Nuclear Munitions Safety Board
NO	Nursing Officer
NOA	Non-Operational Aircraft
NOAA	National Oceanic & Atmospheric Administration (USA)
NOAH	Norwegian Adapted HAWK
NOCUS	North Continental US
NOD	Night Observation Device
NODLR	Night Observation Device, Long Range
NODS	Night Observation & Detection System
NOE	Nap-Of-the-Earth (low flying)
NOGS	Night Observation Gunship system
NOI	Notice of Intention
NOIC	Naval Officer In Charge
NOK	Next Of Kin (UK)
NOL	Naval Ordnance Laboratory (USA)
NOLO	No Live Operator
NOMAD	Naval Operations & Maintenance Aviation Deck (US Navy)
NOMIS	Naval Operations Management Information System
NOP	Office of the Chief of Navy Operations (USA)
NOPT	Naval Organisation Project Team (UK)
NOR	Logic circuit usable as either AND or OR
NORAD	North American Aerospace Defense Command
Noraid	US Organization giving support to Northern Irish republicanism
NORDEC	Nordic Economic Union
NORINCO	China North Industries Corp.
NORLANT	Northern Sub-Area Eastern Atlantic Command (NATO)
NORM	Not Operationally Ready, Maintenance
NORS	Not Operationally Ready, Spare parts
NORSE	Nuclear Optical & Radar Signature Estimation
NORSHIPCO	Norfolk Shipbuilding & Drydock Corp. (USA)
NORTHAG	Northern Army Group (NATO)
NOS	NATO Office of Security
NOS	Night Observation Surveillance sight
NOSC	NATO Operations Support Cell
NOSC	Naval Ocean Systems Center (USA)
NOSIC	Naval Ocean Surveillance Information Center (USA)
NOST	Near-term Optical Sensor Technology
NOTAR	No Tail Rotor
NOVRAM	Non-Volatile Random Access Memory

NPACC	Nominated Primary Alternate Command Centre (NATO)
NPB	Neutral Particle Beam
NPB	Nuclear Powered Bomber
NPBIE	Nuetral Particle Beam Integration Experiment
NPC	NATO Programming Centre
NPE	Navy Preliminary Evaluation (USA)
NPG	Nuclear Planning Group (NATO)
NPLO	NATO Production & Logistics Organisation
NPO	Design & manufacturing facility (CIS)
NPR	No Power Recovery
NPRM	Notice of Proposed Rule Making
NPS	Network Photo System
NPT	Non-Proliferation Treaty
NR	Nuclear Reporting
NR	Submersible research vehicle (USA)
NRA	Nuclear Reaction Analysis
NRC	Nuclear Regulatory Commission (USA)
NRC	Nuclear Reporting Cell
NRDS	Nuclear Rocket Development Station
NRF	Naval Reserve Force (USA)
NRL	Naval Research Laboratory (USA)
NRO	National Reconnaissance Office (USA)
NROSS	Navy Remote Ocean Sensing System / Satellite (US Navy)
NRRC	Naval Reserve Readiness Command (USA)
NRT	Net Registered Tons
NRTS	Not Repairable This Station (USAF)
NRU	North Reference Unit
NRZ	Non-Return to Zero
NSA	National Security Agency (USA)
NSA	National Standards Association (USA)
NSA	New Shipborne Aircraft programme (Canada)
NSA/CSS	National Security Agency, Central Security Service (USA)
NSC	National Security Council
NSC	NATO Supply Centre
NSCO	Naval Shipping Control Officer
NSE	National logistics Support Element (NATO)
NSEA	Naval Sea systems command headquarters (USA)
NSG	NATO Standardisation Group
NSG	Naval Security Group (USA)
NSG	North Seeking Gyrocompass
NSGCHQ	Naval Security Group Command Headquarters (USA)
NSI	Not Seriously Injured
NSIA	National Security Industrial Association (USA)
NSL	Northern Scientific Laboratory (USA)
NSN	NATO Stock Number
NSNF	Non-Strategic Nuclear Forces
NSO	Navigation / Systems Operator
NSP	Navy Special Projects office (USA)
NSR	NATO Staff Requirement
NSR	Naval Staff Requirement (UK)

NSSC	Naval Ship Systems Command (USA)
NSSL	National Severe Storms Laboratory (USA)
NSSMS	NATO Seasparrow Surface Missile System
NST	NATO Staff Target
NST	Naval Staff Target
NST	Non Standard Transmission
NST	Nuclear & Space Talks
NSTL	National Space Technology Laboratories (USA)
NSTN	Naval Shore Telecommunications Network
NSVN	NATO Secure Voice Network
NSWC	Naval Surface Warfare / Weapons Center (USA)
NSWP	Non-Soviet Warsaw Pact
NTA	Not in Target Area
NTAS	NORAD Tactical Autovon System (USAF)
NTAS	Norwegian Tracking Adjunct System
NTB	National Test Bed
NTC	National Territorial Command (Netherlands)
NTC	National Training Center (USA)
NTCCS	Naval Tactical Command & Control System
NTCT	Naval Tactics & Command Trainer
NTDS	Naval Tactical Data System (USA)
NTDS	Naval Tactical Display System
NTDS	Naval Tactical Distribution System
NTF	National Test Facility
NTFWTC	NATO Tactical Fighter & Weapons Training Centre
NTI	National Technology Initiative (USA)
NTIS	Navy Thermal Imaging System
NTM	National Technical Means of evaluation
NTS	Naval Telecommunications System
NTS	Navigation Technology Satellite (US Navy)
NTS	Night Targeting System
NTSA	Navy Tactical Support Activity office (USA)
NTTC	National Technology Transfer Center (NASA) (USA)
NTU	Navigation Training Unit
NTU	New Threat Upgrade
NTWS	New Threat Warning System
NUDETS	Nuclear Detonation Detection & reporting System (NATO)
NUFAS	NATO UHF Frequency Assignment System
NUGP	Nominal Unit Ground Pressure
NV	Night Vision
NV	Normal Vetting (UK)
NV DISTRINAL	NV Distribution International (Belgium)
NVA	Nationale Volksarmee. Former East German Army (Germany)
NVE	Night Vision Equipment
NVG	Night Vision Goggles
NVIS	Nearly Vertical Incidence Skywave
NVS	Night Vision System
NW	Nuclear Warfare
NW	Nuclear Weapon
NWA	Nuclear Weapon Accident

NWARC	Navy Weapons Assesssment Research Centre (China)
NWASI	Northrop Worldwide Aircraft Services Inc. (USA)
NWC	National War College (USA)
NWC	Naval Weapons Center (USA)
NWDS	Navigation & Weapon Delivery System
NWEF	Naval Weapons Evaluation Facility (USA)
NWES	Naval Weapons Engineering Support Activity office (USA)
NWI	Northwest Industries Ltd (Canada)
NWS	North Warning System (USA)
NWSSG	Nuclear Weapon System Safety Group
NZDF	New Zealand Defence Force
NZDSU(SEA)	New Zealand Defence Support Unit (South East Asia)
n/a	Not Applicable

O

OAB	Outer Air Battle
OADS	Omnidirectional Air-Data System
OAF	OAF Graf & Stift AG (Austria)
OAMCE	Organisation Africaine et Malgache de Co-operation Economique
OAMP	Optical Airborne Measurement Platform
OAPEC	Organisation of Arab Petroleum Exporting Countries
OAR	Office of Aerospace Research (USAF)
OAS	Offensive Air Support
OAS	Offensive Avionics System
OAS	On Active Service
OAS	Organization of American States
OASD	Office of the Assistant Secretary of Defence
OASIS	Operational Applications of Special Intelligence Systems (USA)
OAT	Operational Air Traffic
OAT	Outside Air Temperature
OATS	Oxford Air Training School (UK)
OAU	Organisation of African Unity
OB	Ordnance Board
OBC	Optical Barrel Camera
OBE	Off-Board Expendables
OBEWS	On-Board Electronic Warfare Simulator
OBFM	Offensive Basic Flight Manoeuvres
OBI	Omni-Bearing Indicator
OBOGS	On-Board Oxygen Generating System
OBR	Optical Beam Riding
OBS	Observation Balloon System
OBS	On-Board Subsystem
OBU	OSIS Baseline Upgrade
OBVACT	On-Board Visual Aimer Continuation Trainer
OC	Observer Computer
OC	Officer Commanding
OCA	Offensive Counter-Air
OCA	Operational Control Authority
OCAM	Organisation Commune Africaine et Mauricienne
OCC	Obus a Charge Creuse. Shaped charge shell (France)
OCC	Operational Control Centre
OCC	Operator Control Console
OCCIS	Operations Control & Command Information System (USA)
OCCM	Optical Counter Countermeasures
OCCULT	Optical Covert Communications Using Laser Transceivers (USA)
OCD	Central Office for General Defense (Switzerland)
OCEANAV	Oceanographer of the Navy (US Navy)
OCEANLANT	Ocean Sub-Area, Atlantic (NATO)
OCF	Officiating Chaplain to the Forces
OCM	Optical Countermeasures
OCNR	Office of the Chief of Naval Research (USA)
OCP	Operational Computer Program (USA)
OCR	Operational Control Room

OCS	Operational Control Segment
OCTU	Officer Cadet Training Unit
OCU	Operational Conversion Unit (training)
OCU	Operator's Control Unit
OCWP	Operational Control Work Post
OC&S	Ordnance Center & School
OD	Ordinary Seaman
OD	Ordnance Data
OD	Ordnance Delivery
ODA	Official Development Assistance
ODC	Overseas Development Corp. (USA)
ODE	Ordnance Development & Engineering (Singapore)
ODECA	Organisacion de Estados Centro-americanos. Organization of Central American States
ODESSA	Oceanographic Data & Environmental Satellite System Application
ODG	O'Donnell Griffin (Australia)
ODG	Ontario Drive & Gear Ltd (Canada)
ODIN	Operational Data Interface
ODM	Optical Driver Modem
ODR	Overland Downlook Radar
ODS	Optical Disk System
ODT	Omni-Directional Transmission
OE	Opto-Electronic
OEC	Optic-Electronic Corp. (USA)
OECD	Organisation for Economic Co-operation & Development
OED	Operational Evaluation Demonstration
OEEC	Organisation for European Economic Co-operation. Now OECD
OEI	One Engine Inoperative
OEM	Original Equipment Manufacturer
OEP	Office of Emergency Preparedness (USA)
OER	Operational Effectiveness Rate
OETC	Opto-Electronic Technology Consortium (USA)
OEU	Operational Evaluation Unit (Royal Air Force)
OEW	Ordnance & Explosive Waste
OFAB	Oskolochnaya Fugasnaya Aviabomba. Fragmentation demolition bomb (CIS)
OFADJ	Federal Office of Adjudancy (Switzerland)
OFART	Federal Office of Artillery (Switzerland)
OFC	Overhead Foxhole Cover
OFDE	Optronique-Farbricantes de Electronicas Lda (Portugal)
OFINF	Federal Office of Infantry (Switzerland)
OFM	Otto Fuel Monitor
OFP	Operational Flight Programme
OFPA	Federal Office of Armament Production (Switzerland)
OFPP	Office of Procurement & Policy
OFS	Operational Flying School (Royal Navy)
OFSA	Optical Fire Sensor Assembly
OFSAN	Federal Office of Military Health Service (Switzerland)
OFT	Operational Flight Trainer
OFTML	Federal Office of Armoured & Light Troops (Switzerland)

OFTS	Operational Flight & Tactics Simulator
OFTT	Federal Office of Transport Troops (Switzerland)
OFTT	Operational Flight & Tactics Trainer
OG	Observation Group (USA)
OGA	Office General de l'Air (France)
OGE	Out of Ground Effect. Helicopter hovering far above nearest surface
OGMA	Oficinas Gerais de Material Aeronautico (Portugal)
OHC	Overhead Cam Shaft
OHU	Optical Head Unit
OHV	Overhead Valve
OIC	Organisation of the Islamic Conference
OICR	Operational Intelligence Collection Requirement
OIDT	Operator Interactive Display Terminal
OIER	Operational Information Exchange Requirement
OI-N	Office of Information – Navy (US Navy)
OIRA	Official Irish Republican Army
OIU	Operator Interface Unit
OJCS	Office of the Joint Chiefs of Staff
OJT	On the Job Training (USA)
OKB	Missile design bureau (CIS)
OLADE	Organizacion Latinoamericana de Energia
OLAS	Organizacion Latinoamericana de Solidaridad
OLBR	Operational Laser-Beam Recorder
OLOS	Out of Line-Of-Sight
OLTU	Optical Line Terminating Unit
OM	Operations Module
OME	Operational Mission Environment
OMI	Omni-bearing Magnetic Indicator
OMI	Opto Mechanik Inc. (USA)
OMMS	Oxygen Mask-Mounted Sight
OMO	One Man Operated
OMS	Optronic Mast Sensor
OMT	Other Military Targets
OND	Optical Neural Device
ONR	Office of Naval Research (USA)
ONS	Omega Navigation System
OOA	Out Of Area
OOB	Order Of Battle
OOD/W	Officer Of the Day / Watch (UK)
OOV	Objects Of Verification
OP	Observation Post
OPA	Isopropyl alcohol
OPANAL	Organismo para la Prescription de las Armas Nucleares en America Latina
OPBAT	Operations Bahamas And Turks (USCG)
OPCOM	Operational Command
OPCON	Operational Control
OPDEF	Operational Defect
OPDEFSYS	Operational Defects System

OPDIR	Operational Directive
OPEC	Organization of the Petroleum Exporting Countries
OPEVAL	Operational Evaluation
OPFOR	Opposing Forces
OPI	Operator Interface
OPLE	Omega Position-Locating Equipment
OPM	Organisation Papua Merduka
OPN	Optimised-Profile Navigation
OPNOTE	Operational Note
OPORD	Operations Order
OPR	Office of Primary Responsibility
Ops	Operations Officer
OPS DIV	Operations Division, SHAPE (NATO)
OPSTAT	Operational Status
OPV	Observation Post Vehicle
OPV	Offshore Patrol Vessel
OR	Operational Requirement (UK)
OR	Operational Research
OR	Other Ranks
ORACLE	Operational Research And Critical-Link Evaluation
ORADS	Optical Ranging And Detecting Systems (US Army)
ORAE	Aerospace Industries Association of Canada
ORATMS	Off-Route Anti-Tank Mine System
ORB	Omni Radio Beacon
ORB	Operations Record Book
ORBAT	Order Of Battle
ORCC	Operational Requirements Committee (Brunei)
ORCHIDEE	Observatoire Radar Coherent Heliporte d'Investigation Des Elements Ennemis (France)
ord	Ordnance
ORD	Operational Requirement Document
Org	Organisation
ORS	Offensive Radar System
ORT	Optical Relay Tube
ORU	Operator Radio Unit
OS	Observation Squadron
OS	Ordinary Seaman
OSA	Operational Support Aircraft
OSA	Oscilloquartz SA (Switzerland)
OSAF	Office of the Secretary of the Air Force (USA)
OSC	Operational System Control
OSC	Orbital Sciences Corp. (USA)
OSCAD	Office of the Scientific Advisor to SACEUR, SHAPE (NATO)
OSCAR	Optical Submarine Communications by Aerospace Relay (USA)
OSCO	Omori Seikoki Co. Ltd (Japan)
OSD	Office of the Secretary of Defence
OSE	Operations Support Equipment
OSF	Open Systems Foundation
OSF	Optronic Sector System (France)
OSI	On-Site Inspection

OSI	Open Systems Interconnection
OSIS	Ocean Surveillance Information System
OSM	Optical Support Measures
OSN	Office of the Secretary of the Navy (USA)
OSP	Ocean Surveillance Product
OSP	Optical Surveillance Platform
OSPAAAL	Organizacion de Solidaridad con los Pueblos de Africa, Asia e America Latina
OSPDS	Ocean Surveillance Product Dissemination Service
OSS	Office of Strategic Services
OST	Outer Space Treaty
OSU	Omega VLF Sensor Unit
OSV	Ocean Station Vessel
OSV	OPFOR Surrogate Vehicle
OT	Operational Test
OTA	Overflight Top Attack
OTC	Officer in Tactical Command
OTC	Officer Training Corps (UK)
OTC	Operational Training Course (USAF)
OTCIX	Officer in Tactical Command Information Exchange Subsystem
OTD	Operator Training Device
OTE	Operator Training Equipment
OTEA	Operational Test & Evaluation Agency (USA)
OTEC	Operational Test & Evaluation Command (US Army)
OTEVFOR	Operational Test & Evaluation For Operational Requirements (US Navy)
OTH	Over-The-Horizon
OTH-B	Over-The-Horizon, Backscatter
OTHDT	Over-The-Horizon Detection & Targeting
OTHR	Over-The-Horizon Radar
OTH-SW	Over-The-Horizon, Surface Wave
OTH-T	Over-The-Horizon Targeting
OTIS	Observer Thermal Imaging System
OTK	One-Time Key
OTP	On Top Position
OTPI	On Top Position Indicator
OTR	Operational Turn-Round
OTS	Off-The-Shelf
OTS	One-man Operated Ticketing System
OTT	Operations Team Trainer
OTT	Operator Tactics Trainer
OTU	Odometer Transducer Unit
OTU	Operational Training Unit
OTWD	Out-The-Window Display
OU	Operations Unit
OVT	Oceonics Vehicle Technology
OWE	Operating Weight Empty
OWI	Office of War Information
OWL/D	Optical Warning, Location & Detection
OWR	Odenwald Werke Rittersbach (Germany)

OWRM	Other War Reserve Materiel (USA)
OWS	Obstacle Warning System
OWS	Overhead Weapon Station
O&M	Operations & Maintenance
O&O	Operational & Organisational
O/C	On Completion

P

P	Phosphorus
P	Precision
P	Prohibited area
P How	Pack Howitzer
P in C	Paymaster in Chief (UK)
P2S	Panoramic Periscope System
P3I	Pre-Planned Production Improvements
PA	Pilot's Associate
PA	Power Amplifier
PA	Public Address
PAA	Pont Automoteur d'Accompagnement (France)
PAAT	Passive Acoustics Analysis Trainer
PABX	Private Automatic Branch Exchange
PAC	Pacific Aerospace Corp. Ltd (New Zealand)
PAC	Pakistan Aeronautical Complex
PAC	Penetration Aid Carrier
PAC	Programmable Armament Control
PACAF	Pacific Air Forces
PACC	Primary Alternative Command Centre (NATO)
PACCS	Post Attack Command & Control System (USA)
PAD	Packet Assembler / Disassembler. Communications networks
PAD	Planning Analysis Document
PAD	Point Air Defence
PAD	Preliminary Analysis Document (NATO)
PADS	Passive Advanced Sonobuoy
PADS	Position Attack Defence System
PADS	Position & Azimuth Determining System
PADT	Point Air Defence Trainer
PAE	Park Air Electronics Ltd (UK)
PAEI	Plane Avionic Enterprises Inc. (Canada)
PAF	Philippine Air Force
PAF	Pilotage Aerodynamique Fort. Aerodynamic missile flight control (France)
PAHO	Pan-American Health Organization
PAK	Projector Alignment Kit
PAL	Permissive Action Link
PAL	Phase Alternation Line
PAL	Portable Advanced Laser
PALS	Portable Airfield Light Set (USAF)
PALS	Positioning And Locating System
PAM	Penetration Augmented Munition
PAM	Pulse Amplitude Modulation
PAMS	Point Anti-Missile System
PAN	Porte Avions Nucleaire (France)
PANA	Pan-African Press Agency
PAO	Principle Administrative Officer
PAO	Public Affairs Officer (USA)
PAOCCS 2	Portuguese Air Command & Control Systems 2

PAP	Plans Activation Party (NATO)
PAPI	Precision Approach Path Indicator
PAPS	Periodic Armaments Planning System (NATO)
PAPS	Phased Armaments Programme Systems
PAR	Phased Array Radar
PAR	Power Analyser & Recorder
PAR	Precision Approach Radar
PAR	Progressive Aircraft Rework (USA)
PAR	Pulse Acquisition Radar
PARCS	Perimeter Acquisition Radar Characterisation System
PARES	Passive Radar ESM System
PARM	Panzerabwehr-Richtmine (Germany)
PARM	Persistent Anti-Radiation Missile
PARMOD	Progressive Aircraft Rework Modification (USA)
PARS	Primary Attitude Reference Systems
PAS	Performance Advisory System
PAS	Police Aviation Services Ltd (UK)
PAS	Primary Alerting System
PAS	Programmable Audio Synthesiser
PAS	Pulse Analysis System
PASARS	Podded Advanced Synthetic Aperture Radar System
PASS	Passive Aircraft Surveillance System
PASS	Passive & Active Sensor Subsystem
PAT	Power-Assisted Traverse
PATCH	Precision Approach To Coupled Hover
PATE	Pointing And Tracking Experiment
PATHFINDER	Passive Thermal Forward looking Infra-red for Navigation, Detection & Enhanced Resolution
PATHS	Precursor Above-The-Horizon Sensor
PATS	Precision Automated Tracking Station
PATS	Prototype Automatic Target Screener
PAU	Pan-American Union
PAVE-PAWS	Phased-Array Radars
PAWS	Phased-Array Warning System
PB	Particle Beam
PB	Patrol Boat
PB	Plastic Banded
PBAA	Polybutadiene Acrylic Acid
PBAN	Polybutadiene Acrylonitrile
PB-AP	Plastic Banded, Armour Piercing
PB-APDS	Plastic Banded, Armour Piercing, Discarding Sabot
PBASCO	Puritan-Bennett Aero Systems Co. (USA)
PBATS	Portable Battlefield Attack System
PBC	Practice Bomb Carrier
PBDI	Position, Bearing & Distance Indicator
PB-HEPI	Plastic Banded, High Explosive, Penetrating Incendiary
PBI	Peterson Builders Inc. (USA)
PBJ	Partial-Band Jammer
PBN	Pilatus Britten-Norman Ltd (UK)
PBPS	Post-Boost Propulsion System

PBR	Patrol Boat, River (USA)
PbS	Lead sulphide
PB-T	Plastic Banded, Tracer
PBV	Post Boost Vehicle
PBW	Particle-Beam Weapon
PBX	Polymer Bonded Explosive
PBX	Private Branch Exchange
PB(C)	Patrol Boat (Coastal) (USA)
PC	Personal Computer
PC	Printed Circuit
PC	Pulse Compression
PCAS	Persistent Chemical Agent Simulant
PCB	Printed Circuit Board
PCBW	Provisional Combat Bomb Wing (USAF)
PCC	Polarity Coincidence Correlation receiver
PCC	Police Compact Carbine
PCDS	Precision Course Direction System (USA)
PCF	Patrol Craft (Fast) (USA)
PCI	Personal Computer Interface
PCIU	Programmable Communications Interface Unit
PCM	Pulse Code Modulation
PCM	Pyrotechnic Countermeasures
PCP	Personnel Control Point
PCP	Platoon Command Post
PCS	Performance Command System
PCS	Portable Control Station
PCSN	Precision Standard Inc. (USA)
PCTA	Plastic Cased Telescoped Ammunition
PCU	Parachute Control Unit
PCU	Portable Communications Unit
PCU	Power Conditioning Unit
PCU	Power Control Unit
PD	Phenyldichloroarsine. Chemical warfare blister agent
PD	Point Detonating
PD	Principal Directorate (UK)
PD	Programme Director (UK)
PD	Project Definition
PD	Pulse-Doppler (radar)
PDD	Player Detection Device
PDES	Pulse Doppler Elevation Scan
PDF	Popular Defence Force (Sudan)
PDIC	Protection Development International Corp. (USA)
PDM	Pursuit Deterrent Mine
PDMS	Point Defence Missile System
PDNES	Pulse Doppler Non-Elevation Scan
PDP	Passive Driving Periscope
PDRJ	Pulse-Doppler Radar Jammer
PDRM	Portable Dose Rate Meter
PDS	Passive Detection System
PDS	Post Design Support

PDSS Propulsion-Derived Ship Service
PDT Practice Delivery Torpedo
PDTM PD Technical Mouldings plc (UK)
PDU Pilot's Display Unit
PDZC Pathfinder Drop-Zone Control
PE Permanent Echo
PE Plastic Explosive
PE Processing Element
PE Procurement Executive (UK)
PE Protective Entrance
PECM Passive Electronic Countermeasures
PEG Polyethelene Glycol
PEL Portable Event Logger
PELSS Precision Emitter Location Strike System
PENAID Penetration Aid. Countermeasures to help aircraft fly through hostile airspace
PEO Program Executive Officer
PEP Peak Envelope Power
PEPS Positive-Expulsion Propellant System
PERMREP Permanent Representative to NAC (NATO)
PERP Peak Effective Radiated Power
PERT Program Evaluation Review Technique (USA)
PET Polyethylene terephthalate
PETN Pentaerythritetranitrate
PF Preformed Fragment
PFC Parapet Foxhole Cover
PFCS Primary Flight Control System
PFD Primary Flight Display
PFD Proximity Fuze Disconnector
PFF Pathfinder Force
PFF Preformed Fragment(ation)
PFHE Pre-Fragmented High Explosive
PFIB Perfluoroisobutene
PFLP Popular Front for the Liberation of Palestine
PFLP-GC Popular Front for the Liberation of Palestine – General Command
PFLP-SC Popular Front for the Liberation of Palestine – Special Command
PFM Pont Flottant Motorise (France)
PFMGO Pre-Flight Message-Generating Officer (Royal Air Force)
PFN Pulse Forming Network. Electric pulse for ignition
PFOM Preformed Fragmentation OTO Munition
PFP Proximity Fuze Programmer
PFPPX Pre-Fragmented Programmable Proximity Fuzed
PFPX Pre-Fragmented Proximity Fuzed
PFS Primary Flying Squadron
PFT Prefabricate Foxhole Twin
PFTC Plesetsk Flight Test Centre (CIS)
PFTS Permanent Field Training Site
PG Patrol Combatant ship (USA)
PGA Pressure-Garment Assembly
PGCT Precision Gunnery Crew Trainer

PGH	Patrol Gunboat (Hydrofoil) (USA)
PGM	Precision Ground Map radar mode
PGM	Precision Guided Munitions
PGM	Precision Ground Mapping
PGRV	Precision-Guided Re-entry Vehicle (USA)
PGS	Platoon Gunnery Simulator
PGSM	Precision Guided Submunition
PH	Probability of Hit
PHA	Preliminary Hazard Analysis
PHEI	Penetration High Explosive Incendiary
PHI	Position & Heading (or Homing) Indicator
PHIGS	Programmer's Hierarchical Interactive Graphics System
PHM	Patrol Combatant Missile ship (USA)
PHOTINT	Photographic Intelligence
PHQ	Peacetime Headquarters
PHQ	Port Headquarters
PHS	Precision Hover Sensor
PI	Performance Indicator
PI	Photo-Interpreter
PI	Physics International Co. (USA)
PI	Point of Interception
PI	Practice Interception
PI	Process (or Program) Instruction
PI	Product Improvement
PIAF	Piege a Fibre Optique. Ground-based optical fibre detection system (France)
PIBD	Point-Initiating, Base Detonating
PIC	Pilot In Command
PID	Passive / Personal Identification Device
PIDC	Projects International Development Corp. (USA)
PIDP	Programmable Indicator Data Processor
PIE	Pyrotechnically Initiated Explosive
PIF	Photo-Interpretation Facility
PIF	Pilotage en Force. Missile manoeuvrability (France)
PIG	Pendulous Integrating Gyro
PIHM	Protective Integrated Hood Mask
PIJ	Palestinian Islamic Jihad
PILOT	Pod Integrated Localisation, Observation, Transmission
PIM	Position of Intended Movement
PIM	Previously Intended Movement
PIN	Personal Identification Number
PIN	Positive-Intrinsic-Negative (semiconductor)
PINE	Passive Infra-Red Equipment
PINS	Precise Integrated Navigation System
PIO	Public Information Officer (UK)
PIP	Product Improvement Programme
PIRA	Provisional Irish Republican Army
PIRATE	Passive Infra-Red Airborne Track Equipment
PIRAZ	Positive-Identification & Radar Advisory Zone
PIROS	Passive Infra-red Remote Observation System

PISA	Pilot's Infra-red Sighting Ability
PISCES	Plessey Integrated Shipborne Combat Electronic Systems (UK)
PIT	Prioritized Image Transmission
PITS	Passive Identification & Targeting System
PIU	Processor Interface Unit
PIVADS	Product Improved Vulcan Air Defense System (USA)
PJBD	Permanent Joint Board on Defense, Canada-US
PJH	PLRS/JTIDS Hybrid (USA)
PK	Kill Probability
PKP	Predicted Kill Point
PL	Powered Lift
PLA	Palestine Liberation Army
PLA	People's Liberation Army (China)
PLA	Plain Language Address
PLAAF	People's Liberation Army Air Force (China)
PLAN	Operational Plan
PLAN	People's Liberation Army / Navy (China)
PLARS	Position Location & Reporting System (USA)
PLAT	Pilot Landing Aid Television
PLCE	Personal Load Carrying Equipment
PLD	Personal Laser Detector
PLD	Precision Laser Designator
PLF	Popular Liberation Front
PLF	Powered-Lift Facility
PLL	Phase Locked Loop
PLL	Prescribed Load List
PLO	Palestine Liberation Organisation
PLP	Packet Level Protocol
PLRS	Position Location Reporting System
PLS	Palletised Loading System (USA)
PLS	Personal Location System
PLS	Projectile Location System
PLSS	Portable Life-Support System
PLSS	Precision Location Strike System (USA)
PLST	Palletized Loading System Trailer
Plt	Platoon
Plt Off	Pilot Officer
PLYMCHAN	Plymouth Sub-Area Command Headquarters (UK)
PM	Phase Modulation
PM	Plastic Manoeuvre
PM	Porte Mortier. Mortar carrier (France)
PM	Project Manager (UK)
PM	Provost-Marshall
PMA	Projected Map Assembly
PMAS	Performance Measurement Analysis System (USA)
PMCC	Platoon Mistral Command Centre
PME	Professional Military Education (USA)
PMEL	Precision-Measurement Equipment Laboratory (USA)
PMF	Processeur Militaire Francais. Military computer (France)
PMG	Procurement Management Group (UK)

PMH	Patrol Missile Hydrofoil (US Navy)
PML	Pressings (Metal) Ltd (UK)
PML-TNO	TNO Prins Maurits Laboratory (Netherlands)
PMO	Principal Medical Officer
PMO	Program Management Office (USA)
PMP	Pomtommo Mostovoy Park. Pontoon bridge set (CIS)
PMP	Pretoria Metal Pressings (Pty) Ltd (South Africa)
PMR	Pacific Missile Range
PMR	Pritsepniy Minniy Zagraditel. Towed minelayer (CIS)
PMRAFNS	Princess Mary's Royal Air Force Nursing Service
PMRF	Pacific Missile Range Facility (USA)
PMRT	Program Management Responsibility Transfer (USA)
PMS	Pedestal Mounted Stinger
PMS	Personnel Management Squadron (Royal Air Force)
PMS	Portable Monitoring Set
PMTC	Pacific Missile Test Center (USA)
PMTRADE	Program Manager for Training Devices (US Army)
PMTS	Passive Multi-target Tracking System
PMWS	Pedestal-Mounted Weapons System
PNA	Point of No Alternative
PNCS	Performance & Navigation Computer System
PNDC	Provisional National Defense Council (Ghana)
PNET	Peaceful Nuclear Explosions Treaty
PNG	Passive Night-vision Goggles
PNGDF	Papua New Guinea Defence Force
PNGE	Plan Nacional de Guerra Electronica (Spain)
PNJ	Pulsed Noise Jamming
PNO	Principal Naval Overseer (UK)
PNO	Principal Nursing Officer
PNR	Point of No Return
PNST	Pilot Network System Test
PNU	Puma Navigation Update Programme (Royal Air Force)
PNVS	Pilot Night Vision System
PO	Petty Officer (Royal Navy)
PO1/2	Petty Officer 1st/2nd Class (US Navy)
POA	Position Of Advantage
PoACCS Ph II	Portuguese Air Command & Control System Phase II
POC	Point Of Contact
Podas	Portable Data-Acquisition System
PODS	Portable Data Store
PODS	Portable Digitiser Subsystem
POE	Probability Of Error
POET	Primed Oscillator Expendable Transponder
POF	Pakistan Ordnance Factories
POI	Probability Of Intercept
POISE	Pointing & Stabilisation platform Experiment (US Army)
POL	Petroleum, Oil & Lubricants
POL	Policy Division, SHAPE (NATO)
POLAD	Political Adviser
POLISARIO	Frente Popular para la Liberacion de Sabguia el Hamra y Oro

POM	Program Objective Memorandum (USA)
POMCUS	Prepositioning of Organisational Material Configured in Unit Sets (NATO)
POMINS	Portable Mine Neutralisation System
POP	Pipeline Outfit, Petroleum
Port	Left side (looking forward)
POS	Peacetime Operating Stocks
POS	Postes Optiques de Surveillance (France)
POSIX	Portable Operating System Interface for Computer Environments
POST	Passive Optical Seeker Technology
POST	Portable Optical Sensor Testbed
POW	Prisoner Of War
p-p	Peak to peak
PPA	Pre-Planned Attack
PPAS	Precision Pulse Analysis System
PPB&G	Positive Pressure Breathing & G System
PPC	Pulse Position Code (radar uplink / downlink)
PPCM	Perimeter-Portal Continuous Monitoring
PPD	Proximity / Point Detonating
PPI	Pacific Propeller Inc. (USA)
PPI	Plan Position Indicator
PPM	Parts Per Million
PPM	Pulse Power Module
PPS	Photovoltaic Power System
PPS	Precision Pointing System
PPS-SM	Precise Positioning Service – Security Module
PPT	Perspective-Pole Track
PPW	Personal Protection Weapon
PPW	Polizei Pistole Walther (Germany)
PQE	Path Quality Evaluation
PQO	Principal Quality Officer (UK)
PR	Photo Reconnaissance
PRAC	Practice
PRAC-T	Practice Tracer
PRADS	Parachute / Retrorocket Air Drop System
PRAL	Patrulla de Reconocimiento de Alcance Largo. Long-range reconnaissance patrol (El Salvador)
PRB	Pouderies Reunies de Bourges (France)
PRD	Primary Radar Data
PRDS	Processed Radar Display System
PREPSYS	Preparedness for Operations Status System
PRESS	Pacific Range Electromagnetic Signature Study (USAF)
PRESSURS	Pre-Strike Surveillance / Reconnaissance System
PRF	Plug Representing Fuze
PRF	Pulse Repetition Frequency
PRI	Pacific Resources Inc. (USA)
PRI	Projector Reticle Image
PRI	Pulse Repetition Interval
PRISMA	Primary Imaging Systems for Multiple Applications
PRL	Package Research Laboratory (USA)

PRL Programming Research Ltd (UK)
PrNK Pritseino-Navigatsionniy Kompleks. Advanced integrated navigation / attack system (CIS)
PRO Anti-rocket defence (CIS)
PROGSYS Programming System
PROM Programmable Read-Only Memory
PROP Preservation of the Rights of Prisoners
PROSAB Air-launched cluster weapon (CIS)
Prox Proximity
PRP Performance Recovery Program (USA)
PRP Podvizhny Razvedyvatel'ny Punkt. Mobile reconnaissance post (CIS)
PRPA Projectile Raye a Propulsion Additionelle (France)
PRR Production Readiness Review
PRTL Pantser Rups Tegen Luchtdoelen. Anti-aircraft gun system (Netherlands)
PRU Performance Reference Unit
PRW Passive Radar Warning
PS Chloropicrin
PSA Prefabricated Surface, Aluminium
PSC Principal Subordinate Commander (NATO)
PSCS Photographic Sensor Control System
PSD Propulsion System Demonstrator
PSDN Packet Switched Digital Network
PSF Popular Struggle Front (for liberation of Palestine)
PSF Provisional Sinn Fein
PSI Pathfinder Systems Inc. (USA)
PSI Permanent Staff Instructor (UK)
PSI Photon Science Instruments (France)
PSIS Permanent Secretaries Committee on the Intelligence Services
PSK Phase Shift Keying
PSLV Polar Satellite Launch Vehicle
PSMC Pre-Selected Manual Control
PSN Packet Switch Network
PSNR Podvizhnaya, Stantsiya Nazemnoy Razvedki. Mobile ground reconnaissance (battlefield surveillance) radar (CIS)
PSO Pilot Systems Officer (USAF)
PSP Personal Survival Pack
PSP Pointed, Soft Point
PSP Programmable Signal Processor
PSR Passive Ranging Sonar (Germany)
PSR Post-Strike Reconnaissance
PSR Primary Surveillance Radar
PST Pacific Standard Time
PST Propulsion Systems Trainer
PSTAR Portable Search & Target Acquisition Radar
PSTN Public Switched Telephone Network
PSU Passive Sonar (Germany)
PSU Power Supply Unit
PSU Power Switching Unit
PSWHQ Primary Subordinate / Static War Headquarters (NATO)

PSYOPS	Psychological Operations
PS&SSU	Power Supply & System Selector Unit
PT	Penetrant
PT	Plastic Training
PTA	Pilotless Target Aircraft
PTAG	Portable Tactical Aircraft Guidance
PTAS	Pilotless Target Aircraft Squadron
PTCS	Portable Tracking & Control Station
Pte	Private
PTF	Fast Patrol Craft (USA)
PTFE	Polytetrafluoroethene
PTI	Physical Training Instructor
Ptl	Patrol
PTO	Power Take-Off
PTP	Programmable Touch Panel
PTS	Pilot Training Squadron (USA)
PTT	Part Task Trainer
PTT	Post Telegraph & Telephone administration
PTT	Press / Push-To-Talk
PTU	Air pressure, temperature & humidity
PUFFS	Passive Underwater Fire control system
PUMICE	Precision Un-Manned Integrated Control Equipment
PUP	Pick Up Point
PUS	Permanent Under Secretary of State (UK)
PUT	Pop-Up Test of missiles
PV	Patrol Vessel
PV	Positive Vetting (UK)
pvc	Polyvinyl chloride
PVD	Para-visual Director
PVD	Plan View Display
PVO Strany	Voiska Protivovzdushnoy Strany. National air defence command (CIS)
Pvt	Private
PW	Prisoner of War
PW	Pulse Width
PWHQ	Primary War Headquarters
PWI-SR(GR)	Panser Wagen Infanterie-Standaard (Groep)
PWO	Principal Warfare Officer (UK)
PWORC-E	Pratt & Whitney Overhaul & Repair Center-Europe (Netherlands)
PWP	Plasticised White Phosphorous
PWR	Passive Warning Radar
PWR	Pressurised Water Reactor
PWRR	Princess of Wales Royal Regiment (formerly Queens & Hampshires) (UK)
PWRS	Pre-positioned War Reserve Stocks
PWS	Proximity Warning System
PXR	Post Exercise Report (UK)
PYRKAL	Greek Powder & Cartridge Co. SA
PZRK	Perenosniy Zenitniy Raketniy Kompleks. Portable air defence system (CIS)

P&R	Planning & Research Ltd (Israel)
P&SS	Provost & Security Services (UK)
P&WC	Pratt & Whitney Canada Inc.

Q

QA	Quality Assurance
QAAF	Qatari Amiri Air Force
QAF	Qatar Armed Forces
QAO	Quality Assurance Office, Naval Weapons (USA)
QARANC	Queen Alexandra's Royal Army Nursing Corps
QARNNS	Queen Alexandra's Royal Naval Nursing Service
QCB	Quick Change Barrel
QCPSK	Quadrature Coherent Phase Shift Keying
QD	Quantity Distance
QDE	Quality Data Evaluation
QE	Quadrant Elevation
QEC	Quadrantal Error Correction
QF	Quick Firing
QFI	Qualified Flying Instructor (Royal Air Force)
QFM	Quantised Frequency Modulation (USA)
QM	Quartermaster (UK)
QMG	Quartermaster General
QMS	Quality Management System
QMS	Quartermaster-Sergeant
QMSS	Qubit Mine Surveillance System
QOS	Quality Of Service
QPS	Quality Power Supplies Ltd (Israel)
QPSK	Quadrature Phase Shift Keying
QR	Quiet Radar
QR (RN)	Queens Regulations (Royal Navy)
QRA	Quick-Reaction Alert (Royal Air Force)
QRA	Quiet Reconnaissance Airplane (US Army)
QRC	Quick Reaction Capability
QRF	Quick Reaction Force
QRTOL	Quiet RTOL
QSA	Qatar Sea Arm
QSR	Quick Strike Reconnaissance
QSTOL	Quiet STOL
QVI	Quasi-Vertical Incidence sounder
QWI	Qualified Weapons Instructor

R

RA	Rear Admiral
RA	Resolution Advisory
RA	Rocket Assisted
RA	Royal Artillery (UK)
RAAC	Royal Australian Armoured Corps
RAAF	Royal Australian Air Force
RAAMS	Remote Anti-Armor Mine System (USA)
RAAWS	Radar Altimeter & Altitude Warning System
RABDART	Rapid Aircraft Battle-Damage Augmentation Repair Team
RABFAC	Radar Beacon, Forward Air Control
RAC	Radar d'Alert et de Co-ordination (France)
RAC	Radar-Absorbing Chaff
RAC	Royal Armoured Corps
RACE	Rapid Area Clearance Equipment
RACEWS	Racal Automated Communications Electronic Warfare System
RACHEL	Reperage Acoustique d'Helicopteres. Helicopter detection system (France)
RACHID	Reperage Acoustique de Camions et Helicopteres Instrus dans le Desert. Intruder detection system (France)
RACO	Radar Video Converter
RAD	Radiation Accumulated Dose
RADAG	Radar Aimpoint & Guidance
RADAR	Radio Detection And Range (USA)
RADC	Rome Air Development Center (USA)
RADC	Royal Army Dental Corps
RADEX	Rapid Deployment Exercise
RADHAZ	Radiation Hazard
RADIAC	Radioactivity Detection, Indication And Computation
RADIC	Rapidly Deployable Integrated Command & control system
RADIL	ROCC / AWACS Digital Information Link (USA)
RADIOMAR	Hispano Radio Maritima SA (Spain)
RADIRS	Rapid Deployment Multiple Rocket System (USA)
RADM	Rear Admiral (US Navy)
RADOP	Radar Optical fire control system (France)
RADS	Racal Action Display System
RADSIM	Radar Simulator
RAE	Royal Aerospace Establishment. Now part of DRA (UK)
RAE	Royal Australian Engineers
RAEC	Royal Army Educational Corps
RAF	Royal Air Force (UK)
RAF ADAT	Royal Air Force Air Defence Advisory Team
RAFG	Royal Air Force Germany
RAFO	Royal Air Force of Oman
RAFSC	Royal Air Force Support Command
RAFSEE	Royal Air Force Signals Engineering Establishment
RAFTS	Racal Aid For Team Sweeping. Minesweeping system
RAFTS	Reconnaissance Attack Fighter Training System (USA)

RAFVR	Royal Air Force Volunteer Reserve
RAG	Regimental Artillery Group
RAG	Replacement Air Group (US Navy)
RAG	Ring Airfoil Grenade
RAI	Registro Aeronautico Italiano
RAID	Rapid Alerting & Identification Display
RAIDS	Radar Airborne Intrusion Detection System
RAJ	Ring-Around Jammer
RAL	Rutherford Appleton Laboratory (UK)
RALACS	Radar Altimeter Low Altitude Control System
RALS	RAM Alternate Launching System
RAM	Radar Absorbent Material
RAM	Raid-Assessment Mode for radar
RAM	Random Access Memory
RAM	Rolling Airframe Missile
RAMBS	Rapid Anti-personnel Minefield Breaching System
RAMC	Royal Army Medical Corps
RAM-D	Reliability, Availability, Maintainability & Durability
RAMDI	Radioactive Miss-Distance Indicator
RAMP	Rapid Acquisition of Manufactured Parts (US Navy)
RAMP-V	Rapid Multi-Purpose Vehicle
RAMS	Racal Avionics Management System
RAMS	Rapid Assembly of Munitions System
RAMS	Remote Automatic Multipurpose Station
RAMSES	Reprogrammable Advanced Multi-mode Shipborne ECM System
RAMU	Removable Auxiliary Memory Unit
RAN	Royal Australian Navy
RANSAC	Range Surveillance Aircraft
RAN/RAWS	Royal Australian Navy / Role Adaptable Weapons System
RAO	Rear Area Operations
RAO	Royal Army of Oman
RAOC	Regional Air Operations Centre
RAOC	Royal Army Ordnance Corps (now part of Adjutant General's Corps)
RAP	Radar-Absorbent Paint
RAP	Radio Access Point
RAP	Random Access Point
RAP	Recognised Air Picture
RAP	Regimental Aid Post
RAP	Reliable Acoustic Path
RAP	Rocket-Assisted Projectile
RAPC	Royal Army Pay Corps
RAPHAEL TH	Radar de Photographie Aerienne Electronique – Transmission Hertzienne (France)
RAPID	Retrorocket-Assisted Parachute Inflight Delivery
RAPIDS	Radar Passive Identification System
RAPIR	Telepointeur d'Artillerie Portable Infra-Rouge (France)
RAPP	Recognised Air Picture Production
RAPPORT	Rapid Alert Programmed Power Management of Radar Targets
RAPS	Radar Protection System
RAPS	RPV Advanced Payload System (USAF)

RARDE Royal Armament Research & Development Establishment (UK)
RARF Radar, Antenna & RF integrated system
RAS Radar-Absorbing Structures
RAS Rectified Airspeed. IAS corrected for position error
RAS Remote Active Spectrometer
RAS Replenishment At Sea
RASAT Radar And Sonar Alignment Target
RASCAL Radar di Scoperta e Controllo Aereo Locale. Short-range air surveillance radar system (Italy)
RASE Rapid Automatic Sweep Equipment
RASI Radar & navigational aid simulator (Germany)
RASP Rapid Acquisition Spectrum processor
RASP Recognised Air & Surface Picture
RASS Rapid Area Supply Support (USAF)
RAST Radar-Augmented Sub-Target
RASUR Radio Surveillance for intelligence purposes
RASV Reusable Aerospace Vehicle
RAT Ram-Air Turbine
RATAC Radar de Tir pour l'Artillerie de Campagne. Radar for Field Artillery Fire (France)
RATG Ram-Air Turbine Generator
RATO Rocket-Assisted Take-Off
RATS Ranger Anti-Tank System
RATS Rapid Area Transportation Support (USAF)
RATSCAT Radar Target Scatter facility
RATT Radioteletype
RAU Radio Access Unit
RAuxAF Royal Auxiliary Air Force
RAV Robotic Air Vehicle
RAVC Royal Army Veterinary Corps
RAW Rifleman's Assault Weapon
RAWS Radar Altitude Warning System
RAWS Radar Attack & Warning System
Rb Robot (missile prefix) (Sweden)
RB Rapid Bloom
RB Rebated rim
RB Reduced Blast
RB Rifle Brigade
RB Rotating Bolt
RBAF Royal Brunei Armed Forces
RBC Rapid Bloom Chaff / Countermeasures
RBD Radar Beacon Digitizer
RBDF Royal Bahamas Defence Force
RBEB Ribbon Bridge Erection Boat
RBF Remove Before Flight
RBM Rifleman's Breaching Munition
RBOC Rapid Bloom Offboard Countermeasures
RBS Radar Bomb Scoring
RBTS Royal Brunei Technical Services
RBU Raketnaya Bombometnaya Ustanovka. Anti-submarine Rocket Launcher (CIS)

RB/ER	Reduced Blast / Enhanced Radiation
RC	Red Cross
RC	Reserve Component
RC	Rounds Counter
RCAAS	Remote-Controlled Anti-Armor System (USA)
RCAF	Royal Canadian Air Force. (Now part of the CAF)
RCATT	Rapier Classroom Acquisition & Tracking Trainer
RCC	Regional Control Centre
RCC	Resources Control Console for ELINT & COMINT systems
RCC	Robotic Command Centre
RCC	Roland Co-ordination Centre (USA)
RCCN	Reporting / Control Communications Network
RCD	Middle East Regional Co-operation for Development
RCDS	Royal College of Defence Studies (UK)
RCDU	Remote Controlled Defence Unit
RCEC	Remotely-Controlled Explosive Cutter
RCI	Rousseau Controls Inc. (Canada)
RCL	Recoilless
RCM	Radio Countermeasures
RCM	Regimental Court-Martial
RCMAT	Radio-Controlled Miniature Aerial Target
RCMDS	Remote Control Mine Disposal System
RCMG	Rifle-Calibre Machine Gun
RCMP	Royal Canadian Mounted Police
RCMU	Remote Control & Monitoring Unit
RCN	Royal Canadian Navy. (Now Part of the CAF)
RCP	Regimental Command Post
RCS	Radar Cross-Section
RCS	Range Control Station
RCS	Royal Corps of Signals (UK)
RCT	Royal Corps of Transport (UK)
RCTC	Rear Crew Trainer Cabin
RCU	Remote Control Unit
RCV	Remotely Controlled Vehicle
RCV	Robotic Command Vehicle
RC/ATV	Remote Control / All-Terrain Vehicle
RC/MC	Rounds Counter / Magazine Controller
RDA	Remote Display / Alarm
RDAF	Kongelige Danske Flyvevabnet. Royal Danish Air Force
RDAS	Reconnaissance Data Annotation Set
RDAU	Remote Data Acquisition Unit
RDBMS	Relational Database Management System
RDD	Routine Dynamic Display
RDF	Rapid Deployment Force
RDF/LT	Rapid Deployment Force / Light Tank (USA)
RDGT	Reliability Development & Growth Test
RDI	Radar Doppler a Impulsions. Airborne radar (France)
RDJTF	Rapid Deployment Joint Task Force (USA)
RDM	De Rotterdamsche Droogdok Maatschappij BV (Netherlands)

RDM	Radar Doppler Multifunction. Airborne radar
RDM	Remote Demolition Mine
RDMSS	Remotely Deployable Mission Support System
RDP	Radar Data Processor
RDPS	Radar Data Processing System
RDQA	Regional Director, Quality Assurance (UK)
RDS	River Don Stampings Ltd (UK)
RDSS	Rapidly Deployable Surveillance System
RDT	Radartronic A/S (Denmark)
RDT&E	Research, Development, Test & Evaluation (USA)
RDU	Remote Display Unit
RD&A	Research Development & Acquisition
RE	Radar Echo
RE	Research Engineers Ltd (UK)
RE	Research Establishment
RE	Royal Engineers (UK)
REACT	Rain Echo Attenuation Compensation Technology
REACT	Rapid Execution And Combat Targeting program (USA)
READS	Re-Entry Air-Data System
Rear Adm	Rear Admiral
REAS	Realistic Electromagnetic Arena Simulator
ReB	Re-entry Body
REC	Radio-Electronic Combat (CIS)
Recon	Reconnaissance
REDAR	Range Engineering Data Acquisition & Reduction
REDCAP	Real-time Electromagnetic Digitally Controlled Analyser Processor
REDCOM	Readiness Command
ref	Reference
REFORGER	Return of Forces to Germany. Exercise (NATO)
REJAC	Receiver / Jammer Capability
RELP	Residual Excited Linear Prediction algorithm
RELVEL	Relative Velocity
REM	Rocket Engine Module
REMBASS	Remotely Monitored Battlefield Sensor System (USA)
REMDEG	Reliable Military Data Exchange Guide
REME	Royal Electrical & Mechanical Engineers (UK)
REMRO	Remote Radar Operator
REMS	Remote Sensor
REMSEVS	Remote-Control Secure Voice System
REPCO	Reparation Committee of the Big Powers
REPORTER	Radar Equipment Providing Omnidirectional Reporting of Targets at Extended Range (Netherlands)
RER	Radar Electrical Rack (radar uplink / downlink)
RES	Radar Emitter Simulator
RES	Radio Emission Surveillance
RESCAP	Rescue Combat Air Patrol
RESS	Radar Environmental Simulator System
REST	Radar Electronic Scan Technique
RETS	Remote Target System
Reutech	Reunert Technology Systems (Pty) Ltd (South Africa)

REVEAL	Real-time Video Enhancement At Long range
REVISE	Research Vehicle for In-Flight Submunition Ejection programme (UK)
REW	Radio-Electronic Warfare
REWTS	Responsive Electronic Warfare Training System
RF	Radio Frequency
RF	Regional Forces
RFA	Royal Fleet Auxiliary (UK)
RFC	PT Radio Frequency Communication (Indonesia)
RFCT	Rapid Fire Crossing Target
RFDIU	RF / Digital Interface Unit
RFDU	Radio Frequency Distribution Unit
RFI	Radio Frequency Interferometer
RFI	Request For Information
RFIS	Receiver Fire-control computer Interface Software
RFL	Rocket Flare Launcher
RFM	Radio Frequency Module
RFMF	Royal Fiji Military Forces. Army
RFP	Remote Front Panel
RFP	Request for Proposals
RFS	Radio Frequency Surveillance
RFS	Return Fire Simulator
RFT	Rapid Fire Target
RFTR	Ring-Fin Tail Rotor
RFU	Radio Frequency Unit
RG	Reconnaissance Group
RGB	Rail Garrison Basing
RGB	Red, Green, Blue
RGG	Royal Grenadier Guards (UK)
RGJ	Royal Green Jackets (UK)
RGM	Ship-launched surface attack missile (USA)
RGN	Ruchnaya Granata Nastupatel'naya (CIS)
RGO	Ruchnaya Granata Oboronitel'naya (CIS)
RGPO	Range Gate Pull-Off
RGS	Recovery Guidance System
Rgt	Regiment
RGT	Reliability Growth Test
RGWS	Radar-Guided Weapon System
RH	Relative Humidity
RHA	Rolled Homogeneous Armour
RHA	Royal Horse Artillery (UK)
RHAW	Radar Homing And Warning
RHC	Red Hand Commandos (UK)
RHD	Right-Hand Drive
RHF	Royal Highland Fusiliers (UK)
RHG	Royal Horse Guards (UK)
RHI	Range Height Indicator
RHINO	Repeating Hand-held Improved Non-rifled Ordnance
RHKAAF	Royal Hong Kong Auxiliary Air Force
RHWR	Radar Homing & Warning Receiver
RHWS	Radar Homing & Warning System

RI	Risk Index
RIB	Rigid Inflatable Boat
RIC	Reconnaissance Interpretation Centre
RIC	Royal Irish Constabulary
RID	Ridotti, reduced (Italy)
RIDE	Radio Communications Intercept & D/F Equipment
RIDS	Radio Information Distribution System
RIF	Reduction In Force
RIG	Rate Integrating Gyro
RIM	Ship launched surface-to-air missile (USA)
RIMS	Replacement Inertial Measurement System
RIMS	Rockwell Instrumented Millimeter Wave System (USA)
RIN	Royal Institute of Navigation
RINS	Racal Integrated Navigation System
RINT	Radiation Intelligence
RIO	Radar Intercept Officer
RIOT	Remotely Independently Operable Transceiver
RIPP	Radar Information Processing Post
RIRP	Retractable Inflight Refuelling Probe
RIS	Radar Information Service
RIS	Range Information System
RIS	Range Instrumentation System
RIS	Reconnaissance Interface System
RISC	Reduced Instruction Set Computer
RISC	Regional International Security Conference
RISE	Reliability Improved Selected Equipment
RISLS	Receiver Interrogation Sidelobe Suppression
RISS	Remote Infra-red Surveillance System (USA)
RIT	Radio Interface Terminal
RIT	Remote Image Transceiver system
RITA	Reseau Integre de Transmission Automatique (France)
RITAD	Raggrupamento delle Industrie a Tecnologia Avanzata per la Difesa (Italy)
RIU	Radio Interface Unit
RIV	Rapid Intervention Vehicle
RL	Rocket Launcher
RLD	Rijksluchtvaartdienst. Civil aviation department (Netherlands)
RLG	Ring Laser Gyro
RLGM	Remote Loop Group Multiplexer
RLNS	Ring Laser Gyro Land Navigation System
RLT	Rolling Liquid Transporter
RM	Royal Marines (UK)
RMA	Ron McIntosh & Associates Inc. (USA)
RMAS	Royal Military Academy Sandhurst (UK)
RMC	Remote Multiplexer Combiner
RMC	Rocket Motor Case
RMC	Royal Military College (UK)
RMCS	Royal Military College of Science (UK)
RME	Relay Mirror Experiment
RMEF	Reconnaissance Mobile Exploitation Facility

RMG	Ranging Machine Gun
RMI	Remote Magnetic Indicator
RMID	Road Mine Detector
RMP	Recognised Maritime Picture
RMP	Reprogrammable Micro-Processor
RMP	Royal Military Police (Corps of) (UK)
RMR	Remote Map Reader
RMR	Roke Manor Research Ltd (UK)
RMR	Royal Marine Reserve (UK)
RMS	Range Measurement System
RMS	Reconnaissance Management System
RMS	Remote Monitoring System
RMS	Root Mean Square
RMTL	Remorque Militaire de Transport Logistique (France)
RMV	Remotely Manned Vehicle
RMVD	Refresh Memory / Video Generator
RN	Recovery Net
RN	Regency Net. Survivable HF radio C3 system (USA)
RN	Round Nose
RN	Royal Navy (UK)
RNAD	Royal Navy Armament Depot (UK)
RNAS	Royal Naval Air Service (UK)
RNAS	Royal Naval Air Station (UK)
RNAV	Area Navigation
RNAY	Royal Naval Aircraft Yard (UK)
RNC	Royal Navy College (UK)
RNDZ	Rendezvous
RNEE	Royal Navy Equipment Exhibition
RNEFTS	Royal Navy Elementary Flying Training Squadron (UK)
RNF	Radio Navigation Facilities
RNLA	Koninklijke Landmacht. Royal Netherlands Army
RNLAF	Koninklijke Luchtmacht. Royal Netherlands Air Force
RNLMC	Royal Netherlands Marine Corps
RNLN	Koninklijke Marine. Royal Netherlands Navy
RNOAF	Royal Norwegian Air Force
RNR	Royal Naval Reserve (UK)
RNSTS	Royal Naval Supply & Transport Service (UK)
RNTDS	Restructured NTDS
RNVR	Royal Naval Volunteer Reserve (UK)
RNVU	Royal Navy Vetting Unit (UK)
RNXS	Royal Naval Auxiliary Service (UK)
RNZAF	Royal New Zealand Air Force
RNZN	Royal New Zealand Navy
RO	Radio Operator
RO	Range Only type sonobuoy
RO	Receive Only
RO	Royal Ordnance (UK)
RO VIS	Royal Ordnance Vehicle Intercom System
ROAD	Retired On Active Duty
ROBAT	Robotic Counter-Obstacle Vehicle (USA)

ROC	Rate Of Climb
ROC	Regional Operating Centre
ROC	Required Operational Capabilities
ROC	Royal Observer Corps (UK)
ROCA	Republic of China Army
ROCAF	Republic of China Air Force
ROCC	Region Operations Control Centre
ROCN	Republic of China Navy
ROD	Rate Of Descent
RODEO	Radar d'Observation et de Designation d'Objectifs (France)
RODS	Ruggedised Optical Data System
ROE	Rules Of Engagement
ROEREQ	Rules Of Engagement Request
ROF	Rate Of Fire
ROFIS	Royal Ordnance Flash Indiction Shell
Ro-Flow	A ship able to embark smaller craft in a dock
ROKA	Republic of Korea Army
ROKAF	Republic of Korea Air Force
ROKN	Republic of Korea Navy
ROM	Read-Only Memory
ROMAG	Remote Map Generator
ROMANS	Rapid Operational Minefield Attack & Neutralisation System
RON	Research Octane Number
ROPS	Rollover Protective Structure
ROR	Range Only Radar
Ro-Ro	Roll-on Roll-off (ferry)
RORSAT	Radar Ocean Surveillance Satellite
ROSAR	Rotor-SAR (Germany)
ROSIM	Royal Ordnance Simulation (UK)
ROSM	Royal Ordnance Speciality Metals Ltd (UK)
ROST	Radar Operator Training System
ROTA	Royal Ordnance Training Ammunition
ROTHR	Relocatable Over-The-Horizon Radar (USA)
ROV	Remotely Operated Vehicle
ROWPU	Reverse Osmosis Water Purification Unit
RP	Red Phosphorus
RP	Reporting Post
RP	Rocket Propelled
RPA	Radiation Protection Advisor
RPA	Remotely Piloted Aircraft
RPB	Retarded Practice Bomb
RPD	Rapid Inertial Alignment
RPE	Rocket Propulsion Establishment (UK)
RPG	Receiver Processor Group
RPG	Rocket Propelled Grenade
RPG	Rounds Per Gun
RPGT	Radar Procedures Ground Trainer
RPH	Remotely Piloted Helicopter
RPM	Remotely Piloted Munition
RPMB	Remotely Piloted Mini-Blimp

RPOADS Remotely Piloted Observation Aircraft Designator System (US Army)
RPP Resonant Pipe Projector. Acoustic projector
RPRV Remotely Piloted Research Vehicle
RPS Radar Position Symbol
RPS Rikspolisstyrelsen (Sweden)
RPU Receiver Processor Unit
RPV Remotely Piloted Vehicle
RR Recoilless Rifle
RRA Rolls-Royce & Associates Ltd (UK)
RRB Reply Receiver 'B'. Shipborne & ground-based transponder
RRC Rapid Reaction Corps (NATO)
RRE Royal Radar Establishment (UK)
RRF Rapid Reaction Force (NATO)
RRF Ready Reaction Force (New Zealand)
RRF Ready Reserve Force
RRP Rapid Reinforcement Plan (NATO)
RRR Rapid Runway Repair
RRTR Reduced Range Training Round
RRTS Remote Radar Tracking System
RRU Remote Read-Out Unit
RS Radio Set
RS Receiver Segment
RSAF Republic of Singapore Air Force
RSAF Royal Saudi Air Force
RSAF Royal Small Arms Factory (now closed down) (UK)
RSB Rescue & Security Boat
RSC Radar Scattering Camouflage
RSCAAL Remote Sensing Chemical Agent Alarm
RSDU Radar / Sonar Display Unit
RSGS Regional Signal Group, SHAPE (NATO)
RSH Reservoirs Souples Heliportable (France)
RSI Radiation Sciences Inc. (USA)
RSI Radiation Systems Inc. (USA)
RSI Rationalisation, Standardisation, Interoperability
RSIP Radar System Improvement Program (USA)
RSIU Radio Set Interface Unit
RSL Range Safety Launch (Royal Air Force)
RSL Reshef Systems Ltd (Israel)
RSLF Royal Saudi Land Forces
RSLS Receiver Sidelobe Suppression
RSM Regimental Sergeant-Major
RSME Royal School of Military Engineering (UK)
RSN Republic of Singapore Navy
RSNF Royal Saudi Naval Forces
RSO Reconnaissance Systems Officer / Operator
RSP Radar Signal Processor
RSP Ringshell projector. Underwater sound projector
RSR En-Route Surveillance Radar
RSR Ready Service Ring
RSRE Royal Signals & Radar Establishment (UK)

RSS Remote Slave Set
RSS Remote Surveillance System
RSS Rosette Scanning Seeker
RSTA Reconnaissance, Surveillance & Target Acquisition
RSVP Rotating Surveillance Vehicle Platform
RT Reaction Time
RT Remote Terminal
RT Rough Terrain
RTA Receiver / Transmitter Antenna
RTA Rotation Target Altitude
RTAD Remote Trigger Activator Device
RTADS Royal Thai Air Defence System
RTAF Royal Thai Air Force
RTB Return To Base
RTCA Radio Technical Commission for Aeronautics (USA)
RTCH Rough Terrain Container Handler
RTCST Rough Terrain Container Straddle Truck
RTD Radar Target Designator control
RTD Rontgen Technische Dienst BV (Netherlands)
RTDT Real-Time Data Transfer
RTE Rail Transfer Equipment
RTF Remote Test Facility
RTFL Rough Terrain Forklift
RTG Raketen Technik GmbH (Germany)
RTI Research Triangle Institute (USA)
RTLS Return To Launch Site
RTM Remote Telemetry Module
RTN Royal Thai Navy
RTOL Reduced Take-Off & Landing
RTP Real Time PHIGS
RTP Research & Technology Projects
RTR Real-Time Reconnaissance
RTRCDS Real-Time Reconnaissance Cockpit Display System
RTS Radar Target Simulator
RTS Radar Test Station
RTS Remote Tracking Station
RTS Replacement Training Squadron
RTS Robot Target System
RTSS Remote Transmit Select System
RTT Roues Transporteur de Troupes (France)
RTTC Regional Technology Transfer Center (NASA) (USA)
RTTL Range Target-Towing Launch (Royal Air Force)
RTTS Robotic Target Training System
RTTY Radioteletype
RTY Raw Total Yield, megatons
RUC Royal Ulster Constabulary
RUMS Rechner Unterstutztes Morser System. Mortar co-ordination system (Germany)
RV Radar Vector
RV Reconnaissance Vehicle

RV	Re-entry Vehicle
RV	Rendezvous
RV	Rescue Vessel (Royal Air Force)
RVA	Radar Vectoring Area
RVD	Residual Vapour Detection
RVDP	Radar Video Data Processing
RVDU	Radar Video Distribution Unit
RVI	Renault Vehicules Industriels SA (France)
RV-PVO	Radio Teknicheski Voiska-PVO. Radio technical troops (CIS)
RVR	Rear-View Radar
RVR	Runway Visual Range
RVT	Remote Video Terminal
RV/WH	Re-entry Vehicle / Warhead
RW	Radiological Weapon
RW	Reconnaissance Wing
RWI	Radio Wire Integration
RWR	Radar Warning Receiver
RWR	Rear-Warning Radar
RWS	Radar Warning System
RWT	Radar Warning Trainer
RX	Receive
R&D	Research & Development
R&H	Van Rietschoten & Houwens BV (Netherlands)
R&M	Reliability & Maintainability
R&S	Rohde & Schwarz GmbH & Co.KG (Germany)
R/C	Radio Control
R/EO	Radar / Electro-Optical
R/O	Radio Operator
R/Pt	Reporting Point
R/T	Radio Telephone
R/T	Receiver / Transmitter

S

S of S	Secretary of State for defence (UK)
S3	Serial Signalling System
SA	Arsine. Chemical warfare blood agent
SA	Scientific Adviser (UK)
SA	Simple Alert
SA	Surface-to-Air
SAA	Small Arms & Ammunition
SAAC	Simulator for Air-to-Air Combat
SAAF	Small Arms Alignment Fixture
SAAF	South African Air Force
SAAHS	Stability Augmentation & Attitude Hold System
SAAM	Special-Assignment Airlift Mission
SAAM	Systeme naval d'autodefense moyen portee. Naval surface-to-air missile system (France)
SABA	Small Agile Battlefield Aircraft
SABA	Swiss & British Associates Ltd (UK)
SABAR	Satellites, Balloons And Rockets
SABCA	Societe Anonyme Belge de Constructions Aeronautiques (Belgium)
SABMIS	Surface-to-Air Ballistic Missile Interception System
SABRES	Small Arms Battlefield Realistic Engagement Simulator
SAC	Senior Aircraftsman
SAC	Shaanxi Aircraft Co. (China)
SAC	Shenyang Aircraft Corp. (China)
SAC	Strategic Air Command (USA)
SACADVON	Strategic Air Command Advanced Echelon
SACC	Secondary Alternative Command Centre (NATO)
SACC	Small Area Camouflage Cover
SACCS	Strategic Air Command Automated Command Control System (USA)
SACDIN	SAC Digital Information Network (USA)
SACEUR	Supreme Allied Commander, Europe (NATO)
SACEURREP	SACEUR Representative to the Military Committee (NATO)
SACF	Stand-Alone Control Facility
SACINTNET	Strategic Air Command Intelligence Network (USA)
SACL	Strobe Anti-Collision Light
SACLANCTEN	SACLANT Anti-submarine Warfare Research Centre (NATO)
SACLANT	Supreme Allied Commander, Atlantic (NATO)
SACLOS	Semi-Automatic Command to Line-Of-Sight
SACMFCS	Small Arms Common Module Fire Control System
SACRU	Semi Automatic Cargo Release Unit
SACS	Secondary Attitude & Compass System
SACS	Small Air-Capable Ship
SACS	Speed / Attitude Control System
SACT	Signal Acquisition Conditioning Terminal
SACTIS	Submarine Action Information System (Italy)
SACU	Stand-Alone digital Communications Unit
SAD	Submarine Anomaly Detector
SADA	Semi-automatic Air Defence system (Spain)
SADANG	Systeme Acoustique d'Atlantique Nouvelle Generation (France)

SADARM	Search And Destroy Armor Munition (USA)
SADF	South African Defence Force
SADM	Special Atomic Demolition Munition
SAdO	Senior Administration Officer (Royal Air Force)
SADRAL	Systeme d'Autodefense Rapproche Anti-aerien Leger. Light short-range anti-aircraft self-defence system (France)
SADRAM	Sense / Seek & Destroy Radar-Assisted Mission
SADRG	Semi-Active Doppler Radar Guidance
SADS	Submarine Active Detection System
SADS-TG	Submarine Active Detection System – Transmit Group
SAE	Singapore Automotive Engineering Pte Ltd
SAE	Society of Automotive Engineers
SAF	Semi-Automated Forces
SAF	Singapore Armed Forces. Army
SAFCS	Standard Automatic Flight Control System
SAFF	Safing, Arming, Fuzing, Firing
SAFFIRE	Synthetic Aperture Fully Focused Imaging Radar Equipment
SAFOS	Office of the Secretary of the Air Force (USA)
SAFS	Sonar And Fire Control System
SAFU	Safety, Arming & Fuzing Unit
SAG	Salzburger Aluminium AG (Austria)
SAG	Support Air Group (Royal Navy)
SAG	Surface Action Group (US Navy)
SAG	Survivability Analysis Group (USA)
SAGE	Semi-Automatic Ground Environment (Air Defence)
SAGW	Surface-to-Air Guided Weapon (UK)
SAH	Semi-Active Homing
SAHR	Semi-Active Homing Radar
SAHRS	Standard Attitude & Heading Reference System
SAI	Swedish Aerospace Industries Association
SAIC	Science Applications International Corp. (USA)
SAIF	Standard Avionics Integrated Fuzing
SAIL	Sonar / Artificial Intelligence Laboratory (USA)
SAINCO	Sociedad Anonima de Instalaciones de Control (Spain)
SAINT	Sonar Analysis Initial Trainer
SAINT	Surveillance, Acquisition, Identification, Notification & Tracking
SAIRS	Standardised Advanced Infra-Red Sensor
SAIS	Shipboard Automated Interconnect System
SAIT	SAI Technology (USA)
SAL	Security Access Level
SAL	Selected Altitude Layer decoder
SAL	Semi-Active Laser
SAL	Standard Aoro Ltd (Canada)
SAL	Strategic Arms Limitation
SALS	Shipborne Aircraft Landing System
SALT	Strategic Arms Limitation Talks / Treaty
SAM	Sound-Absorbing Material
SAM	Special Air Missions squadron (USAF)
SAM	Standard Avionic Module
SAM	Surface-to-Air Missile

SAMAN	Service d'Approvisionnement en Materiel de l'Aeronautique Navale (France)
SAMHADS	Submarine Automated Message Handling And Distribution System
SAMM	Societe d'Applications des Machines Motrices (France)
SAMOC	SAM Operation Centre
SAMOS	Satellite And Missile Observation Station
SAMP	Systeme sol-Air Moyenne Portee (France)
SAMP/N	Surface-to-Air Missile Platform / Naval
SAMP/T	Sol-Air Moyeenne Portee / Terrestre. Surface-to-Air Missile Platform / Terrain
SAMS	Small Automatic Message Switch
SAMS	South African Medical Service
SAMSO	Space And Missile Systems Organisation
SAMSON	Special Avionics Mission Strap-On Now (pod)
SAMSON	Strategic Automatic Message Switching Operational Network
SAMT	State-of-the-art Medium Terminals (Canada)
SAMTEC	Space And Missile Test Center (USA)
SA-N	Surface-to-Air (Navy) missile
SAN	South African Navy
SanABw	Sanitatsamt der Bundeswehr (Germany)
SANDAIRE	San Diego Aircraft Engineering Inc. (USA)
SANTAL	Systeme Anti-Aerien Leger. Light anti-aircraft system (France)
SAO	Special Access Only. Security classification
SAOCS	Submarine / Aircraft Optical Communications System (USA)
SAP	Seaborne Air Platform
SAP	Semi-Armour Piercing
SAP	Sensing And Processing unit
SAP	Shijiazhuang Aircraft Plant (China)
SAP	Simulated Attack Profile
SAP	South African Police
SAPHEI	Semi-Armour Piercing, High Explosive, Incendiary
SAPI	Semi-Armour Piercing, Incendiary
SAPIENS	Surveillance Alarm & Protection, Intelligence EW Naval Systems
SAPLIC	Small Arms Projected Line Charge
SAPOM	Semi-Armour-Piercing OTO Munition
SAPOMER	Semi-Armour-Piercing OTO Munition, Extended Range
SAR	Search And Rescue
SAR	Semi-Active Radar
SAR	Short Assault Rifle
SAR	Stand-Alone Radar
SAR	Synthetic Aperture Radar
SAR	Synthetic Array Radar
SARA	Station Aerotransportable de Reconnaissance Aerienne (France)
SARAH	Search And Rescue And Homing
SARBE	Search And Rescue Beacon Equipment (UK)
SARCAP	Search And Rescue Combat Air Patrol
SARG	Semi-Active Radar Homing
SARIE	Selective Automatic Radar Identification Equipment
SARIS	Synthetic-Aperture Radar Interpretation System
SARK	Saville Advanced Remote Keying

SARP	Semi-Automatic Radar Plotting
SARP	Signal Auto Radar-Processing system
SARP	Sistema Avtomatichyeskoy Radiolokatsionnoi Prokladki. System of automatic radiolocation (CIS)
SARS	Secondary Attitude Defence System
SARS	Static Automatic Reporting System
SARSAT	Search And Rescue Satellite Aided Tracking
SART	Self-Activating Reactive Target
SART	Semi-active Artificial Radar Target
SARTS	Self-Activating Reactive Target System
SAR/FTI	Synthetic Aperture Radar / Fixed Target Indicator
SAS	Schnellbrucke auf Stutzen (Germany)
SAS	Second Acoustic Source
SAS	Special Air Service (UK)
SASC	Senate Armed Services Committee (USA)
SASMEX	Safety At Sea International, Maritime Electronics Exhibition
SASO	Senior Air Staff Officer (Royal Air Force)
SASP	Single Advanced Signal Processor
SASRS	Satellite-Aided Search & Rescue System
SASS	Small Aerostat Surveillance System
SASS	Strategic Airborne Surveillance System
SAT	Sea Acceptance Testing
SAT	Situational Awareness Technology
SAT	Small Arms Trainer
SAT	Small Arms Transmitter
SAT	Societe Anonyme de Telecommunications (France)
SAT	Synchronised Acoustic Transmitter
SATAID	Sonar Auto-detection & Tracking, AI & Data fusion
SATAIR	Sea Acceptance Trials, Air
SATC	Ship Acoustic & Torpedo Countermeasures
SATCOM	Satellite Communications
SATCP	Systeme Anti-Aerien a Tres Courte Portee. Very short range anti-aircraft system (France)
SATF	Strike And Terrain-Following radar
SATIN	Strategic Air Command Automated Total Information Network (USA)
SATIN	Survivability Augmentation for Transport Installation – Now
SATIR	System zur Auswertung Taktischer Informationen auf Raketenzerstorern. Tactical data handling system (Germany)
SATKA	Surveillance Acquisition Tracking & Kill Assessment
SATM	Scatterable Anti-tank Mine
SATNAV	Satellite Navigation
SATNET	Satellite Network
SATO	South Atlantic Treaty Organization
SATS	Small Airfield for Tactical Support
SATS	Small Arms Target System
SATSIM	Saturation Countermeasures Simulator (USAF)
SATT	Small Aircraft Training Target
SATURN	Second Generation of Anti-jam Tactical UHF Radios for NATO
SAU	Safety & Arming Unit
SAU	Signal Acquisition Unit

SAU	Surface Attack Unit
SAV	Surface Auxiliary Vessel
SAVA	Standard Army Vectronics Architecture (USA)
SAVAN	Stabilised Aiming, Vertical sensing And Navigation
SAVR	Strapdown Attitude / Velocity Reference
SAW	Special Air Warfare (USAF)
SAW	Squad Automatic Weapon
SAW	Surface Acoustic Wave
SAWE	Simulated Area Weapons Effects
SAWE-NBC	Simulated Area Weapons Effects-Nuclear, Biological & Chemical
SAWE-RF	Simulated Area Weapons Effects, Radio Frequency
SAWE-RF	Small Arms Weapons Effects, Radio Frequency
SAWES	Small Arms Weapons Effect Simulator
SAWHQ	SHAPE Alternative War Headquarters (NATO)
SAWS	Satellite Attack Warning System
SAWS	Silent Attack Warning System
SAWSS	Shipboard Aircraft Weight Subsystem
SA/BM	Systems Analysis & Battle Management
SBA	Spot-Beam Aerial / Antenna
SBAC	The Society of British Aerospace Companies Ltd
SBHe	Space-Based Hypervelocity Experiment
SBHRG	Space-Based Hypervelocity Rail Gun
SBI	Space-Based Interceptor
SBIR	Small Business Innovative Research program (USA)
SBKEWS	Space-Based Kinetic Energy Weapon System
SBKKV	Space-Based Kinetic Kill Vehicle
SBL	Space-Based Laser
SBL/BMD	Space-Based Laser / Ballistic Missile Defense (USA)
SBNPB	Space-Based Neutral Particle Beam
SBP	Sub-Bottom Profiler
SBPB	Space-Based Particle Beam
SBR	Space-Based Radar
SBRC	Santa Barbara Research Center (USA)
SBSS	Standard Base-Supply System (USAF)
SC	Self-Cocking
SC	Single Card
SCA	Section Chief Assembly. Part of gun display unit
SCA	Simulation Control Area
SCAD	Scientific Adviser (NATO)
SCAD	Subsonic Cruise Armed Decoy
SCADC	Standard Central Air Data Computer (USAF)
SCADS	Shipborne Containerised Air Defence System
SCALE	Syllabically Companded And Logically Encoded delta modulation
SCAM	Strike Camera
SCAMBA	Scatterable Minefield Breaching Apparatus
SCAMP	Sperry Computer Aided Message Processor (UK)
SCAR	Sistemi de Control de Armamento (Italy)
SCAR	Strike Control And Reconnaissance
SCARS	Status Control Alert Report System
SCAS	Site Chemical Agent Detection System

SCAT	Scout / Attack Helicopter
SCAT	South China Aero Technology Ltd (Hong Kong)
SCAT	Speed Command of Attitude & Thrust
SCC	Sector Command Centre
SCC	SHAPE Command Centre (NATO)
SCC	Standing Consultative Commission
SCC	System Control Centre
SCD	Semi-Conductor Devices (Israel)
SCD	Signal Command Decoder
SCDL	Surveillance & Control Data Link
SCDU	Smoke Detection Control Unit
SCE	Safety Consulting Engineers Inc. (USA)
SCEPC	Senior Civil Emergency Planning Committee
SCEPS	Stored Chemical Energy Propulsion System
SCG	Self Changing Gears
SCG	Special Consultative Group
SCI	Serial Communication Interface
SCI	Structural Composites Industries (USA)
SCIDA	System Co-ordinating Installation Design Authority
SCIF	Secure Compartmented Information Facilities
SCIMITAR	System for Countering Interdiction Missiles & Target Acquisition Radars
SCIP	Shipboard Countermeasures Interface Package
SCIRP	Semiconductor IR Photography
SCMP	Supreme Council for Military Police (Iran)
SCMS	Submarine Communication Management System
SCNS	Self-Contained Navigation System
SCOMP	Secure Communications Processor
SCONUM	Ship Control Number
SCORE	Stratified Charge Omnivorous Rotary Engine
SCORE	Surface Coastal Radar Equipment
SCOT	Satellite-Communications On-board Terminal
SCOTT	Single Channel Objective Tactical Terminal
SCOTT	Submarine Combat Team Trainer
SCOUT	Short-range Combined Observation, Uncooled Thermal
SCPC	Single Channel Per Carrier
SCPI	Supersonic Cruise Propulsion Integration
SCPS	Survivable Collective Protection System
SCR	Signal-Controlled Rectifier
SCR	Single Channel Radio system
SCRA	Single Channel Radio Access
SCRAM	Supersonic Combustion Ramjet
SCRT	Single Channel Transponder Receiver
SCS	Satellite Communication System
SCS	Sidewinder Control System
SCS	Singapore Computer Systems Limited
SCS	Special Communications System
SCS	Submarine Control Simulator
SCS	Survivable Control System

SCSC	Strategic Conventional Stand-off Capability
SCSG	Satellite Communications Sub-Group
SCSI	Small Computer System Interface
SCT	Single Channel Transponder
SCT	Societe des Ceramiques Techniques (France)
SCTS	Ship's Compass Transmitting System
SCTS	System Components Test Station
SCTT	Submarine Command Team Trainer
SCU	Station Control Unit
SCU	Stores Control Unit
SCU	Supplemental Control Unit
SCU	Switching & Control Unit. Communication management system
SCU	System Control Unit
SCUBA	Self-Contained Underwater Breathing Apparatus
SCUD	Subsonic Cruise Unarmed Decoy
SCV	Single Card Vocoders
SCWA	Single Channel Wire Access
SCWG	Satellite Communications Working Group
SD	Saco Defense Inc. (USA)
SD	Self Destruct
SD	Self-Destroying
SD	Service Dress (UK)
SDA	Strategic Defence Architecture
SDA	System Design Authority
SDAT	Silicon Diode Array Target
SDBA	Short-Duration Breathing Apparatus
SDC	Space Defence Centre
SDC	Strategic Defence Command (USA)
SDC	Supreme Defence Council (Iran)
SDDU	Sight Display Drive Unit
SDF	Self-Destruct Fuze
SDI	Strategic Defense Initiative (USA)
SDIO	Strategic Defense Initiative Organization (USA)
SDIP	Strategic Defense Initiative Program (USA)
SDIPO	Strategic Defense Initiative Participation Office (USA)
SDK	Skin Decontamination Kit
SDM	Scatter-Drop Mine / Munition
SDM	Site-Defense of Minuteman
SDM	Somali Democratic Movement
SDMS	Shipboard Data Multiplex System (USA)
SDMS	Support Defence Missile System
SDNRIU	Secure Digital Net Radio Interface Unit
SDNS	Security Data Network Service
SDP	Signal Data Processor
SDP	Sloboda DP (Yugoslavia)
SDPR	Federal Directorate of Supply & Procurement (Yugoslavia)
SDRS	Splash-Detection Radar System
SDS	Satellite Data System (USA)
SDS	Secondary Data System
SDS	Small Digital Switch

SDS	Steering & Depth Control Simulator
SDS	Strategic Defense System (USA)
SDSMS	Self-Defence Surface Missile System
SDS(E)	S&D Security (Equipment) Ltd (UK)
SDU	Satellite Data Unit
SDU	Secure Data Unit
SDU	Special Diver's Unit
SDV	Swimmer Delivery Vehicle (USA)
SEA	Sachse Engineering Associates Inc. (USA)
SEAC	South-East Asia Command
SEAC	Submarine Exercise Area Controller
SEACASS	Ships Electronic Audio Control And Selection System
SEACOR	Systems Engineering Associates Co. (USA)
SEAD	Suppression of Enemy Air Defence(s)
SEAFAC	Systems Engineering Avionics Facility
SEAL	Sea / Air / Land
SEAPAC	Sea-Activated Parachute Automatic Crew release
SEATO	Southeast Asia Treaty Organisation
SEB	Systems Electronics Box
SEC	Societa Esercizio Cantieri SpA (Italy)
SEC	Spoiler & Elevator Computer
SEC	Submarine Element Co-ordinator
SECCOS	Office of the Secretary to the Chief of Staff, SHAPE (NATO)
SecDef	Secretary of State for Defense (USA)
Sect	Section
SECU	Spoilers Electronic Control Unit
Sec/HQUKLF	Command Secretary / Headquarters, United Kingdom Land Forces
SED	Safe Escape Distance
SED	Sensor Evolutionary Development
SEDIS	Surface-Emitter Detection Identification System
SEE	Small Emplacement Excavator
SEFIS	Simulated Electronic Flight Instruments System
SEFT	Section d'Etudes et de Fabrications des Telecommunications (France)
SEID	Systeme d'Ecartometrie Infra-rouge Differentielle. Infra-red differential missile tracking system (France)
SEISP	System Electronic Information Security Policy
SEL	Alcatel SEL AG (Germany)
SELA	Sistema Economico Latinoamericano
SELCAL	Selective Calling
SEloKaH	System Elektronische Kampffuhrung Heer (Germany)
SEM	Security Engineered Machinery (USA)
SEM	Standard Electronic Module
SEMA	Special Electronic Mission Aircraft
SEMC	Standard Electronic Memory Cartridge
SEME	School of Electrical & Mechanical Engineering
SEMMB	Societe d'Exploitation des Materiels Martin-Baker (France)
SEMSA	Simulateur d'Entrainement a la Maintenance Systeme d'Arme (France)
SEN	Shell Extended Range NORICUM

SEN	Small Extension Node. Communications
SENEGAMBIA	Senegal & Gambia
SENGO	Senior Engineering Officer (Royal Air Force)
SENIT	Systeme d'Exploitation Navale des Informations Tactique (France)
SENS	Small Extension Node Switch. Communications
SENSO	Sensor Operator
SEOW	Single Engineering Order Wire
SEP	Single-Event Phenomenon
SEP	Societe Europeenne de Propulsion (France)
SEP	Spherical Error Probability
SEP	System Executive & Planning
SEPA	Societa di Elettronica per l'Automazione (Italy)
SEPADS	Sonar Environmental Prediction And Display System
SERC	Science & Engineering Research Council (UK)
SES	Secure Equipment System
SES	Software Exploitation Segment (JSIPS)
SES	Surface Effect Ship
SESN	Selenia Elsag Sistemi Navali (Italy)
SESP	Space Experiment Support Program (USAF)
SETAF	Southern European Task Force (NATO)
SETS	Seeker Evaluation Test System
SEU	Sensor Electronics Unit
SEU	Sight Electronics Unit
SEU	Single-Event Upset
SEU	System Electronic Unit
SEV	Specially Equipped Vehicle
SEV	Surface Effect Vehicle
SEVENATAF	Seventh Allied Tactical Air Force (NATO)
SEVVA	Security Evaluation, Validation & Verification Agency (UK)
SEWACO	Sensor Weapon Control And Command system
SEWS	Satellite Early Warning System
SEWS	SIGINT / EW Subsystem (USA)
SEWT	Simulator for Electronic Warfare Training
SEZ	Selector Engagement Zone
SF	Signal Frequency
SF	Supporting Fire
SFA	Super Freinage et Amorage. Super retarded & arming (France)
SFAR	Special Federal Aviation Regulation (USA)
sfc	Specific Fuel Consumption
SFCS	Simplified Fire Control System
SFCS	Submarine Fire Control System
SFCS	Survivable Flight Control System
SFF	Self-Forming Fragment
SFIM	Societe de Fabrication d'Instrument de Mesure (France)
SFIRR	Solid Fuel Integral Rocket / Ramjet
SFM	Self-Forging Munition
SFM	Sensor Fuzed Munitions
SFP	Sensor Fusion Post
SFSO	Station Flight-Safety Officer (Royal Air Force)
SFTS	Service Flying Training School

SFTS	Synthetic Flight Training System
SFU	Societe des Fonderies d'Ussel (France)
SFV	Stinger Fighting Vehicle (USA)
SFW	Sensor Fuzed Weapon
SG	Shell Gun, cannon
SG	Sorties Generated
SG	Sub-Group
SGDCN	Second-Generation Defence Communications Network (New Zealand)
SGDN	Secretariat General de la Defense Nationale (France)
SGEMP	System-Generated EMP
SGLS	Space-Ground Link Subsystem (USAF)
SGML	Standard Generalized Markup Language
SGSSA	Small Ground Station Segmented Antenna
Sgt	Sergeant
SGT	Satellite Ground Terminal
SGU	Signal Generator Unit
SH	Parachute retarded (CIS)
SH	Sary Shagan (CIS)
SH	Schuberth Helme GmbH (Germany)
SH	Squash Head
SHACE	Supreme Headquarters Allied Command Europe (NATO)
SHADL	Shahine Data Link (France)
SHADOW	Subsonic Hovering Armament Direction & Observation Window (UAV)
SHAEF	Supreme Headquarters of the Allied Expiditionary Force
SHAPE	Supreme Headquarters Allied Powers, Europe (NATO)
SHARP	Standard Hardware Acquisition & Reliability Programme
SHARS	Strapdown Heading & Attitude Reference System
SHEER	Signaal HF ECCM Equipment Range (Netherlands)
SHELF	Super Hard ELF System (USA)
SHF	Super High Frequency (3-30 GHz)
SHIELDS	Ships Highpower Electronic Defense System
SHINCOM	Shipboard Integrated interior Communications system
SHINMACS	Shipboard Integrated Machinery Control System
SHINPADS	Shipboard Integrated Processing And Display System (Canada)
SHOC	Supreme Headquarters Operations Centre (NATO)
SHODOP	Short-Range Doppler
SHOM	Service Hydrographique et Oceanographique de la Marine (France)
SHORAD	Short-Range Air Defence
SHORAN	Short Range Navigation
SHORGUT	Short Range Gun Training
SHR	Superheterodyne Receiver
SHS	Ship Handling Simulator
SHUD	Smart Head-Up Display
SHUP	Silo-Hardness Upgrade Program (USAF)
SH/PRAC	Squash Head / Practice
SI	International System of Units
SI	Seriously Injured
SI	Special Intelligence

SIA	Scanner Infra-red Assembly
SIAM	Self-Initiated Anti-aircraft Missile
SIAr	Service de Surveillance Industrielle de l'Armement (France)
SIASS	Submarine Integrated Attack & Surveillance Sonar (Netherlands)
SIB	Security Investigation Branch (UK)
SIB	Submarine Indicator Buoy
SIB	Systems Implementation Branch (NATO)
SIBCA	Sampling & Identification of Biological & Chemical Agents
SIC	Standard Industrial Classification
SIC	Subject Indicator Code
SICAS	Ship Installed Chemical Alarm System
SICBM	Small Intercontinental Ballistic Missile
Sicile	Station Integree de Communications pour Interventions Legeres. Multi-media communication station (France)
SICLAMEN	Systeme d'Identification Compatible L-band l'Ami ou Ennemi (France)
SICM	Small Intercontinental Missile
SICOMORE	Simulateur de Centre Operations Modulaire et Reconfigurable
SICOT	Systeme Individuel de Commandement de Terrain. Personal battlefield command & control system (France)
SICP	Subscriber Interface Computer Program (USA)
SICPS	Standardised Integrated Command Post Shelter
SICS	Ship Installed Chemical System (UK)
SICS	Ships Integrated Communications System (France)
SICS	Submarine Integrated Combat System
SID	Information Service, Ministry of National Defense (Belgium)
SID	Situation Information Display console
SID	Standard Instrument Departure
SIDENT	Site Identification
SIDES	Societe Industrielle pour le Developpement de la Securite (France)
SIDFAC	Shore Integration & Demonstration Facility
SIDS	Secondary Image Dissemination System
SIE	Self-Initiated Elimination
SIF	Selective Identification Facility
SIF	Selective Identification Feature
SIF	Shore Integration Facility
SIF	System Integration Facility (USA)
SIFIDA	Societe Internationale Financiere pour les Investissements et le Development en Afrique
SIFONET	Shipboard Integrated Fibre Optic Network
SIFV	Sensor Instantaneous Field of View
SIG	Signature (USA)
SIGHT	Signaal General-purpose High-resolution Tactical workstation (Netherlands)
SIGINT	Signals Intelligence
SIGMA	Signaal Modular Architecture (Netherlands)
SIGMA	System for Interception, Goniometry, Monitoring & Analysis
SIGSEC	Signal Security
SILAT	Latecoere Societe Industrielle d'Aviation (France)
SILLACS	Siemens Low Level Air defence Control System (Germany)
Sim	Simulator / Simulation

SIM SAM Intercept Missile
SIMBAD Systeme Integre de Mistral Bitube d'Auto-Defense. Close-in surface-to-air missile system (France)
SIMFICS Simulator Fire Control System
SIMOP Simultaneous Operation (of co-located RF sets)
SIMOX Separation by Implantation of Oxygen
SINBADS Submarine Integrated Battle And Data System (Netherlands)
SINCGARS Single Channel Ground / Air Radio System (USA)
SINCOS Signaal Integrated Communications System (Netherlands)
SINS Ship's Inertial Navigation System
SINTAC Systeme Integre d'Identification, de Navigation, de Controle du Trafic Aerien, Anti-collision et de Communication (France)
SIOP Single Integrated Operations Plan
SIP Surface Impact Propulsion
SIP System Improvement Programme
SIPE Soldier Integrated Protective Ensemble (US Army)
SIPF Royal Solomon Islands Police Force
SIPRI Stockholm International Peace Research Institute (Sweden)
SIPS Small Integrated Propulsion System
SIR Search & Interrogation Radar
SIR Shuttle Imaging Radar
SIRA Sensor Infra-Red Acoustic
SIRAS Surveillance & IFF Radar Simulation
SIRCS Shipboard Intermediate-Range Combat System
SIRE Satellite Infra-Red Experiment
SIRPA Service d'Information et de Relations Publiques des Armees (Française)
SIRS Ship Installed Radiac System
SIS Secret Intelligence Service
SIS Super Instrumentation System
SISMAS Sonar In-Situ Mode Assessment System
SISP System Interconnection Security Policy
SIT Silicon Intensified Target
SITCEN Situation Centre
SITO Security Industry Training Organisation Ltd (UK)
SITREP Situation Report
SIU Sensor Interface Unit (USAF)
SIU Sidewinder Interface Unit
SIU Standard Interface Unit
SIVG System Instandsetzungs und Verwertangsgesellschaft (Germany)
SIXATAF Sixth Allied Tactical Air Force (NATO)
SJ Ski-Jump
SJAC The Society of Japanese Aerospace Companies Inc. (Japan)
SJB Semi-Jetborne
SjvS Forsvarets Sjukvardsstyrelse (Sweden)
SKAD Survival Kit Air Droppable
SKE Station Keeping Equipment
SKF Superkritischer Flugel. Supercritical wing (Germany)
SKL Statens Kriminaltekniska Laboratorium (Sweden)
SKOT Sredni Kolowy Opancerzny Transporter. Armoured personnel carrier (Poland)

SKS	Simonov self-loading rifle (CIS)
SKT	Specialty Knowledge Tests (USAF)
SL	Space Launcher
SL	Start Line
SL	Streamlined (boat-tailed)
SLAF	Sri Lankan Air Force
SLAM	Selectable Lightweight Attack Munition
SLAM	Stand-off Land Attack Missile
SLAM	Supersonic Low-Altitude Missile
SLAMMR	Side-Looking Airborne Modulated Multi Mission Radar
SLAMS	Surface Look-Alike Mine Systems
SLAMSHOT	Shot Location And Monitoring, Shows Hit On Target
SLAP	Saboted Light Armor Penetrator
SLAR	Side-Looking Airborne Radar
SLAT	Ship-Launched Air Target
SLAT	Slow, Low, Airborne Target
SLAT	Supersonic Low Altitude Target
SLBM	Sea-Launched Ballistic Missile
SLBM	Submarine Launched Ballistic Missile
SLC	Sonobuoy Launch Container
SLC	Submarine Laser Communications programme (USA)
SLC	Systems & Logistics Corp. (USA)
SLCM	Sea / Ship-Launched Cruise Missile
SLCM	Submarine-Launched Cruise Missile
SLEP	Service Life Extension/Enhancement Program (USA)
SLEP	Slimline Equipment Practice
SLEWS	Shipboard Lightweight EW System
SLFCS	Survivable Low Frequency Communications System (USA)
SLIM	Surface-Launched Interceptor Missile
SLKT	Survivability, Lethality & Key Technologies
SLMM	Sea-Launched Mobile Mine
SLOC	Sea Line Of Communications
SLOS	Stabilised Long-range Observation System
SLOS	Star Line-Of-Sight
SLOTTS	Surface Launch Organic Tactical Training System
SLP	Survivor Locator Package (USAF)
SLR	Side-Looking Radar
SLR	Super Low Recoil
SLRR	Side-Looking Reconnaissance Radar
SLS	Side-Lobe Suppression
SLS	Single Lane System
SLSAR	Side-Looking Synthetic-Aperture Radar
SLt	Sub-Lieutenant
SLU	Stabilised Laser Unit
SLU	Surface Launched Unit
SLUFAE	Surface-Launched Unit, Fuel-Air Explosive
SLUMINE	Surface-Launched Unit, Mine
SLUTT	Ship-Launched Underwater Transponder Target
SLV	Satellite Launch Vehicle

SLV	Space-Launched Vehicle (USAF)
SLWPG	Senior Level Weapons Protection Group (NATO)
SLWT	Side Loading Warping Tug (USA)
SM	Sandeep Metalcraft Pvt Ltd (India)
SM	Sergeant-Major
SM	Simulation Module
SM	Smoke
SM	Special Mission (USA)
SM	Standard Missile
SMA	Scientific Management Associates Pty Ltd (Australia)
SMA	Segnalamento Marittimo ed Aero SpA (Italy)
SMA	Signal Message Address
SMA	Systems Management American Corp. (USA)
SMAC	Scene Matching by Area Correlation
SMADS	Software Maintenance Development Access Switch (NATO)
SMARCENT	Service des Marches Centralises (France)
SMART	Sensor Fuzed Munition for Artillery
SMART	Signaal Multibeam Acquisition Radar for Targeting
SMART	Small Arms Trainer
SMART	Spurt Message Alphanumeric Radio Terminal
SMART-T	Secure Mobile Anti-jam Reliable Tactical Terminal (US Army)
SMAS	Ship Motion & Anemometer Simulator
SMASH	Southeast Asia Multisensor Armament System, Helicopter
SMATCALS	Signature Managed Air Traffic Control Approach & Landing System (US Navy)
SMAW	Shoulder-launched Multi-purpose Assault Weapon
SMAW-D	Shoulder-launched Multi-purpose Assault Weapon, Disposable
SMC	Sioux Manufacturing Corp. (USA)
SMC	South Midlands Communications Ltd (UK)
SMCO	Standard Manufacturing Company
SMCS	Scatterable Mine Clearance System
SMCS	Standard Machinery Control System
SMCS	Submarine Command System
SMD	Surface-Mounted Devices
SMDU	Strapdown Magnetic Detector Unit
SME	Special Mission Equipment
SME	Syarikat Malaysia Explosives Sdn Bhd
SMES	Strategic Missile Evaluation Squadron
SMES	Superconductive Magnetic Energy Storage
SMET	Simulated Mission Endurance Testing
SMG	Societe Metallurgique de Gerzat (France)
SMG	Sub-Machine Gun
SMHMS	Standardised Magnetic Helmet-Mounted Sight
SMI-E	Shipboard Mounting Interface-Electrical
SMILS	Sonobuoy Missile-Impact Location System
SMK	Smoke
SMMR	Scanning Multi-Frequency Microwave Radiometer
Smoke BE	Smoke Base Ejection
Smoke WP	Smoke White Phosphorus
SMP	Stores Management Processor

SMP	Surface Mine Plough (UK)
SMR	Selective Message Routeing
SMR	Specialised Mobile Radio
SMRS	Stores Management & Release System
SMS	Scatterable Mine System
SMS	Sensor Monitoring Set
SMS	SIGINT Manpack System
SMS	Space Mission Simulator
SMS	Stores Management System
SMS	Strategic Missile Squadron
SMS	Supply & Movements Squadron (Royal Air Force)
SMT	Static, Mobile or Transportable
SMT	Surface Mount Technology
SMTC	Space & Missiles Test Center (USAF)
SMTD	STOL / Manoeuvring Technology Demonstrator
SMTI	Selective Moving-Target Indicator
SMTO	Space & Missile Test Organisation (USAF)
SMUD	Stand-off Munitions Disrupter
SMUTS	Submarine-Mounted Underwater Tracking System
SMW	Strategic Missile Wing
Sm/Co	Samarium / cobalt
SNAP	Steerable Null Antenna Processor
SNAP	Synchronous Numeric Array Processor
SNAP	Systems for Nuclear Auxiliary Power
SNAR	Stantsiya Nazemnoy Artilleriyskoy Razvedki. Artillery location radar (CIS)
SNAW	School of Naval Air Warfare (UK)
SNC	Standard Navigation Computer
SNCO	Senior Non-Commissioned Officer (UK)
SNDV	Strategic Nuclear Delivery Vehicle
SNEP-C	Saudi Naval Expansion Programme, Communications
SNF	Short-range Nuclear Force
SNF	Somali National Front
SNFC	Standing Naval Force, Channel (NATO)
SNLC	Senior NATO Logisticians Conference
SNLE	Sous-marines Nucleaire Lanceur d'Engins balistique. Nuclear-powered ballistic missile submarine (France)
SNM	Somalia National Movement
SNM	Special Nuclear Material
SNO	Senior Nursing Officer
SNORT	Supersonic Naval Ordnance Rocket Track
SNPE	Societe Nationale des Poudres Explosifs (France)
SNR	Signal-to-Noise Ratio
SNS	Shared Network Server
SNTI	Systeme Numerise de Transmissions Interieures (France)
SNVS	Stabilised Night-Vision Equipment
SO	Signal Officer
SO	Staff Officer
SO in C	Signal Officer-in-Chief (UK)
SOA	Somalia Democratic Front

SOA	Speed Of Advance
SOARS	Satellite Onboard Attack Warning System
SOAS	Submarine Operational Automation System
SOBIC	Shore-Based Intelligence Centre
SOC	Sector Operations Centre
SOC	Sparton of Canada Ltd
SoCal	Southern California International Consulting Company (USA)
SOCOM	Special Operations Command (USA)
SOCUS	South Continental US
SODA	System for Observation & artillery Data Acquisition
SODEPAX	Committee on Society, Development & Peace
SODERN	Societe Anonyme d'Etudes et Realisations Nucleaires (France)
SOE	Special Operation Executive
SOF	Special Operations Forces
SOF	Stand-Off Flare
SOF	Strategic Offensive Forces
SOFA	Status Of Forces Agreement
SOFAG	Special-Operations Force Assistance Group
SOFAR	Sound Ocean Fixing And Ranging
SOFATS	Special Operations Forces Aircrew Training System
SOG	Special Operations Group (USA)
SOJ	Stand-Off Jamming
SOLAS	Safety Of Life At Sea
SOLCA	Special Operation Liaison Communications Assemblage
SOLIC	Special Operations & Low Intensity Conflict (Symposium & Exhibition)
SOLL	Special Operations, Lo-Level
SON	Statement of Operational Need
SONAR	Sound Navigation And Ranging
SOO	Standard Operational Orders
SOON	Solar-Obsevatory Optical Network (USAF)
SOP	Standard Operational Procedure
SOP	Submarine Operational Programme
SOR	Specific Operational Requirement
SOR	State Of Readiness
SOR	Struck Off Records
SORAO	Sottosistema di Sorveglianza ed Acquisizione Obiettivi (Italy)
SORAS	Sound Ranging System
SORT	Simulated Optical Range Tester
SOS	Save Our Souls. Distress signal
SOS	Special Operations Squadron (USAF)
SOS	Stabilised Optical Sight
SOSB	System Operating & Support Branch (NATO)
SOSUS	Sound Surveillance System
SOTAS	Stand-Off Target Acquisition System
SOTD	Stabilised Optical Tracking Device
SOV	Simulated Operational Vehicle
SOV	Special Operations Vehicle
SOVMEDRON	Soviet Mediterranean Squadron
SOW	Special Operations Wing

SOW	Stand-Off Weapon
SO(S)	Supply Officer (Stores) (UK)
Sp	Support
SP	Self-Propelled
SP	Shore Patrol (USA)
SP	Single-Phase
SP	Spotting round
SP	Stabilised Platform
SPA	Special Purpose Aircraft
SPAAG	Self-Propelled Anti-Aircraft Gun
SPAAM	Self-Propelled Anti-Aircraft Missile
SPACECOM	Space Command (USAF)
SPAD	Signal Processing And Display sonar
SPADATS	Space Detection And Tracking System (USA)
SPADCCS	Space Command And Control System
SPADE	Space Acquisitions Defence Experiment
SPADOC	Space Defense Operations Center (USA)
SPAG	Self-Propelled Assault Gun
SPAL	Simulator Projectile Airburst Liquid
SPANNER	Special Analysis of Net Radio (USA)
SPAR	Solid-state Phased-Array Radar
SPARK	Solid Propellant Advanced Ramjet Kinetic energy missile
SPARTA	Special Anti-missile Research Tests, Australia (USA)
SPASM	Self-Propelled Air-to-Surface Missile
SPASUR	Space Surveillance
SPASYN	Space Synchro
SPAT	Self-Propelled Acoustic Target
SPATG	Self-Propelled Anti-Tank Gun
SPAW	Self-Propelled Artillery Weapon
SPAWAR	Space & Naval Warfare Systems Command (USA)
SPB	Section Personnel Bridge
SPB	Systems Policy Branch (NATO)
SPBA	Silent Partner Body Armor, Inc. (USA)
SPC	Special Projects Co-ordination (UK)
SPC	Stored Program Control
SPC	Synthetic Particulated Chaff
SPCD	Space Communications Division (USAF)
spec	Specification
SPEC	South Pacific Bureau for Economic Co-operation
SPED	Supersonic Planetary Entry Decelerator
SPEES	Systeme Pour l'Elevation de l'Endommagement Structural (France)
SPEL	Sociedade Portuguesa de Explosivos SA
SPER	Strategic Planning Executive Review
SPEW	Small Platform Electronic Warfare system
SPF	Strategic Offensive Forces
SPG	Self-Propelled Gun
SPG	Special Patrol Group
SPH	Self-Propelled Howitzer
SPI	Surface Position Indicator
SPI	Suspension & Parts Industries Ltd (Israel)

SPIDER Signaal Portable Infantry Digital Encrypted Radio (Netherlands)
SPILS Spin Prevention & Incidence Limiting System
SPILS Stall Protection & Incidence-Limiting System
SPINTCOM Special-Intelligence Communications
SPIRE Skeletal Perspective Image
SPIRIT Spectral Infra-red Rocket-borne Interferometer Telescope
SPL Self-Propelled Launcher
SPLA Sudanese People's Liberation Army
SPM Self-Propelled Mortar
SPM Somali Patriotic Movement
SPN/GEANS Standard Precision Navigator / Gimballed Electrically Aircraft Navigation System
SPO Safety Petty Officer
SPO System Program Office (USA)
SPOT Speed, Position, Track
SPP Special Purpose Pistol
SPR Secure Packet Radio
SPR Solid-Propellant Rocket
SPRA Special-Purpose Reconnaissance Aircraft
SPRITE Signal Processing In The Element
SPRITE Surveillance, Patrol, Reconnaissance, Intelligence-gathering, Target-designation & Electronic warfare (UK)
SPRN Special Projects, Royal Navy
SPS Self-Protection Subsystem
SPS Standard Positioning Service
SPSDS Shipboard Passive Surveillance & Detection System
SPSM Sensorgezundete Panzerabwehr Submunition (Germany)
SPT-B Selectable-Performance Target, Ballistic
SPU Samokhodnaya Puskovaya Ustanovka
SPU Signal Processing Unit
SPW Self-Protection Weapon
SQ Super Quick. Fuzes
SQ System Qualifier
SQL Structured Query Language
sqn Squadron
Sqn Ldr Squadron Leader
SQUIRE Signaal Quiet Universal Intruder Recognition Equipment
SR Search Rate
SR Short-Range
SR Solid Rocket
SR Strategic Reconnaissance (USA)
Sr Col Senior Colonel
SRA Shop Repair Assembly
SRA Special Repair Activities
SRA Surveillance & Reconnaissance Aircraft
SRAAM Short Range Air-to-Air Missile
SRAAW Short-Range Anti-Air Warfare
SRAM Short-Range Attack Missile
SRAM Static Random Access Memory
SRARM Short-Range Anti-Radiation Missile

SRAW	Short-Range Anti-tank Weapon
SRB	Solid Rocket Booster
SRBM	Short-Range Ballistic Missile
SRBOC	Super Rapid Blooming Offboard Chaff
SRBS	Skeletal Reference Baseline Simulator, SDI (USA)
SRBT	SRB Technologies Inc. (USA)
SRCR	Short-Range Chaff Rocket
SRCU	Secure Remote Control Unit
SRDE	Search Radar Data Extractor
SRDE	Signals Research & Development Establishment
SRDL	Saunders-Roe Developments Ltd (UK)
SRDP	Sborno-Razbornoye-Dorozhnoye Pokrytiye. Glued plywood roadway sections (CIS)
SRE	Surveillance Radar Element
SREMP	Source Region EMP protection
SRF	Strategic Rocket Forces
SRFLS	Self-Repairing Flight Control System
SRG	Shell Replenishment Gear
SRHIT	Short-Range Hit-to-kill missile
SRHIT	Short-Range Homing Intercept Technology
SRI	Short-Range Insert
SRI	Southwest Research Institute (USA)
SRINF	Short-Range Intermediate Nuclear Forces
SRL	Systems Research Laboratories Inc. (USA)
SRM	Short-Range Missile
SRM	Solid Rocket Motor
SRMH	Single Role Minehunter
SRMP	Short-Range Maritime Patrol
SRO	Second Radio Operator
SRO	Senior Ranking Officer
SROB	Short-Range Omnidirectional Beacon
SRP	Stabilisation Reference Package
SRP/PDS	Stabilisation Reference Package / Position Determining System
SRR	Short-Range Recovery
SRS	Scientific Radio Systems Inc. (USA)
SRS	Sonobuoy Reference System
SRS	Survival Radio Set
SRT	Southern Research Technologies Inc. (USA)
SRT	Standard Remote Terminal (USA)
SRT	System Readiness Test
SRTS	Short-Range Thermal Sight
SRU	Shop-Replacable Unit
SRU	Slip Ring Unit
SR-UAV	Short-Range UAV programme (USA)
SRV	Surrogate Research Vehicle (USA)
SRW	Strategic Reconnaissance Wing (USAF)
SS	Attack Submarine
SS	Search Unit (China)
SS	Shot Sensor
SS	Single Shot

SS	Surface-to-Surface
SS	System Simulation
SSAG	Auxiliary Submarine (USA)
SSAI	Schneider Sohn & Associates Inc. (USA)
SSAN	Auxiliary Nuclear Powered Submarine
SSATDS	Surface Ship Anti-Torpedo Defence System
SSB	Conventional-powered Ballistic Missile Submarine
SSB	Single Sideband
SSBN	Nuclear-Powered Ballistic Missile Submarine
SSBS	Sol-Sol Balistique Strategique. Surface-to-surface ballistic missile (France)
SSC	Solid-state Scanner
SSCP	Sea System Controllerate Publication
SSCPU	Shipboard Sweep Control & Positioning Unit
SSCS	Surface Ship Command System
SSCT	Small Ships' Communications Terminal
SSD	System Security Database
SSD	Systems & Simulation Division (USAF)
SSDE	Submerged Signal & Decoy Ejector
SSDF	Somalia Salvation Democratic Front
SSDS	Ship Self-Defense System (US Navy)
SSDSG	Special State Defense Study group (USA)
SSE	Singapore Shipbuilding & Engineering Ltd
SSE	Submarine Signal Ejector
SSF	Station Services Flight (Royal Air Force)
SSFM	Steerable Sensor-Fuxed Munition
SSFTC	Sary Shagan Flight Test Center (CIS)
SSG	Guided Missile Submarine
SSG	Single Shot Gun
SSGN	Nuclear-Powered Cruise Missile Submarine
SSgt	Staff Sergeant
SSGW	Surface-to-Surface Guided Weapon
SSIP	Subsystem Integration Project (NATO)
SSIXS	Submarine Satellite Information Exchange System
SSK	Conventional Submarine (diesel-electric)
SSK	Single Shot Kill
SSKP	Single Shot Kill Probability
SSL	Single Station Location
SSL	Submarine Safety Lanes
SSM	Sea-Skimmer Missile
SSM	Surface-to-Surface Missile
SS-N	Surface-to-Surface (Navy) missile
SSN	Nuclear-Powered Attack Submarine
SSNP	Syrian Social Nationalist Party
SSNR	Samokhadnaya Stantsiya Navendeniya – Raket. Mobile missile guidance station (CIS)
SSO	System Security Officer
SSP	System Security Policy
SSPA	Solid-State Phased-Array
SSPC	Single Station Command Post
SSPI	Sighting System Passive / IR

SSR	Secondary Surveillance Radar
SSRT	Samokhadnaya Stantsiya Razvedki Tseleukazaniya. Mobile detection & designation radar (CIS)
SSS	Space Surveillance System (US Navy)
SSS	Strategic Satellite System
SSS	System Support Segment (JSIPS)
SSSI	Sikorsky Support Services Inc. (USA)
SST	Single Subscriber Terminal
SST	Solid-State Transmitter
SST	Special Surface Target torpedo
SSTD	Surface Ship Torpedo Defence
SSTI	Stabilised Steerable Thermal Imaging
SSTOL	Super-Short / Supersonic & Short Take-Off & Landing
SSTS	Space Surveillance & Tracking System
SSTV	Satcom Secure Voice Terminal
SSTX	Solid-State Transmitter
SSU	Sensor Surveying Unit
SSURADS	Shipboard Surveillance Radar Systems
SSVC	Services Sound & Vision Corp. (UK)
SSVDT	Sight, Stabilised, Visual Data Transmitting
SSXBT/SSXSV	Submarine-launched Expendable Bathythermograph/Sound Velocity
ST	Strategic Transport, air
ST	Support Tank
ST3S	Service Technique des Systemes Strategiques et Spatiaux (France)
STA	Shell Transfer Arm
STA	Surveillance & Target Acquisition
STACOS	Signaal Tactical Command System (Netherlands)
STADAN	Space Tracking And Data Network
STAFF	Smart Target Activated Fire & Forget
STAFF	Smart Top Attack Fire & Forget
STAG	Strategy & Tactics Analysis Group (US Army)
STAJ	Short-Term Anti-Jam
STALOC	Self-Tracking Automatic Lock-On Circuit
STAMO	Stable Master Oscillator
STAMP	Small Tactical Aerial Mobility Platform (USA)
STANAG	Standard NATO Agreement
STANAVFORCHAN	Standing Naval Force, Channel (NATO)
STANAVFORLANT	Standing Naval Force, Atlantic (NATO)
STANAVFORMED	Standing Naval Force, Mediterranean (NATO)
STANO	Surveillance, Target Acquisition & Night Observation
STANOC	Surveillance, Target Acquisition, Night Observation & Counter Surveillance Centre
STAPL	Ship-Tethered Aerial Platform
STAR	Ship Tactical Airborne RPV
STAR	Standard Terminal Arrival Route
STAR	Strategic & Tactical Airborne Recovery
STAR	Submarine Target
STAR	Surface To Air Recovery
STAR	System Threat Assessment Report

Starboard	Right side (looking forward)
STARCOM	Strategic Army Communications (USA)
STARS	Small Tethered Aerostat Relocatable System
STARS	Surveillance Target Attack Radar System
START	Strategic Arms Reduction Talks / Treaty
STARTEX	Start of Exercise (UK)
STARTLE	Surveillance & Target Acquisition Radar for Tank Location & Engagement (USA)
STAS	Stabilised Thermal Imaging System
STAT	Section Technique de l'Armee de Terre (France)
STB	Super Tropical Bleach
STB V	Support Tournant Bouclier. Shield rotary mount (France)
STC	Satellite Test Centre
STC	Sea Training Centre (Royal Navy)
STC	Sensitivity Time Control
STC	SHAPE Technical Centre (NATO)
STC	Strike Command
STC	Supplemental Type Certificate (USA)
STC	Support Tank Command
STC	Swept Time Constant
STCICS	Strike Command Integrated Communications System (UK)
STD	Standard
STDL	Submarine Tactical Data Link (USA)
STDN	Space-Tracking Data Network
STE	Simulateur de Tir Embarque (France)
STEARS	Stand-off Tactical Electronic Airborne Reconnaissance System
STEI	Service Technique de l'Electronique et de l'Informatique (France)
STEM	System Trainer & Exercise Module
STEMS	Standard Emitter Simulator
STEPS	Simulation, Training, Education & Procedural System
STEVI	Sperry Turbine Engine Vibration Indicator
STE/FVS	Simplified Test Equipment / Fighting Vehicle System
STE/ICE	Simplified Test Equipment / Internal Combustion Engine (USA)
STFV	Sensor Total Field of View
STGR	Search, Track & Guidance Radar
STH	Singapore Technologies Holdings Pte Ltd
STICAR	Systeme de Transmission des Informations Codees des Armees (France)
STICS	Scalable Transportable Communications System
STIR	Separate Tracking & Illumination Radar
STIR	Surveillance Target Indicator Radar
STIRS	Strapdown Inertial Reference System
STIS	Stabilised Thermal Imaging System
STM	Supersonic Tactical Missile
STO	Short Take-Off
STOAL	Short Take-Off & Arrested Landing
STOC	Special Tactical Operations Center (US National Military Command Center)
STOL	Short Take-Off & Landing
STORM	Sector Tactical Operations Range Modules

STORM	Sensored Tactical Off-Road Mine
STORMS	Sense, Tank Off-Route Mine System
STORMS	Standard Stores Management System
STOVL	Short Take-Off / Vertical Landing
stp	Standard temperature & pressure
STP	Sensor Track Processor
STP	Status Test Panel
STP	Systems Technology Programme
STPA	Service Technique des Programmes Aeronautiques (France)
STR	Sidetone Ranging modulation
STR	Support Tournant sur Rail. Rail-mounted rotary mount (France)
STR	Systems Technology Radar
STR TA	Support Tournant sur Rail Tout Azimuth. All-round rail-mounted rotary mount (France)
STRADIS	Structure Analysis Design & Implementation of Information Systems
STRAP	Sonobuoy Thinned Random Array Project (USA)
STRAP	Straight Through Repeater Antenna Program (USA)
STRC	Strategic Training Route Complex
STRIDA	Systeme de Traitement et de Representation des Informations de Defense Aerienne (France)
STRIKFLANT	Strike Fleet Atlantic (NATO)
STRIKFORSOUTH	Naval Strike & Support Forces, Southern Europe (NATO)
STS	Satellite Transmission Systems Inc. (USA)
STS	Support & Test Station
STSMT	Service Technique des Systemes de Missiles Tactiques (France)
STST	Strategic Transportable Satellite Terminal
STT	Single Target Tracking
STTE	Service Technique des Telecommunications et des Equipements Aeronautiques (France)
STUP	Spinning Tubular Projectile
STV	Surrogate Teleoperated Vehicle
STVTS	Submarine Tactical Visual Training System
STW	Special-to-Weapon Equipment
STWA	Short Trailing-Wire Antenna
STWS	Shipborne Torpedo Weapon System
SUAWACS	Soviet Union Airborne Warning & Control System
SUBACLANT	Submarine Allied Command, Atlantic (NATO)
SUBCHECK	Submarine Check message
SUBEASTLANT	Submarine Forces, Eastern Atlantic Command (NATO)
SUBICS	Submarine Integrated Combat Systems
SUBLOOK	Submarine Look
SUBMED	Submarines Mediterranean (NATO)
SUBMISS	Submarine Missing message
SUBNOTE	Submarine Note message
SubOpAuth	Submarine Operating Authority
SUBROC	Submarine-Launched Rocket
SUBSAFE	Submarine Safe message
SUBSUNK	Submarine Sunk message
SUBTACS	Submarine Tactical Communication System
Sub-TDS	Tactical Data System for Submarines (Denmark)

SUBWESTLANT	Submarine Forces, Western Atlantic Command (NATO)
SUCCESSOR	Submarine / Surface Ship Command & Control Evolutionary System for Successive Operational Requirements (UK)
SUCOC	Succession Of Command
SUIT	Sights, Unit, Infantry, Trilux (UK)
SUM	Structural Usage Monitor
SUM	Surface-to-Underwater Missile
SUMB	Simca-Unic Marmon-Bocquet
SUMED	Suez-Mediterranee
SUMMADE	System Universal Modular Mine And Demolition Explosives
SUMS	Shallow Underwater Missile System
SUP	Surface, Practice
SUPT	Specialised Undergraduate Pilot Training (USA)
Surg	Surgeon
SURTASS	Surveillance Towed Array Sensor System
SURVSATCOM	Survivable Satellite Communications (USA)
SUS	Signal, Underwater Sound
SUSAT	Sights, Unit, Small Arms, Trilux (UK)
SUSV	Small Unit Support Vehicle
SUT	Surface & Underwater Target
SUU	Secondary User Unit
SUU	Suspended Underwing Unit
SUWN-1	Surface to Underwater Missile Launcher (CIS)
SVA	Security Violation Alert
SVA	Societe Vendomoise d'Avionique (France)
SVD	Dragunov sniper rifle (CIS)
SVO	Special Operations Vehicle (Land Rover) (UK)
SVS	Secure Voice Switch
SVSE	Secure Voice Switch Equipment
SVTT	Surface Vessel Torpedo Tube
SV.AM	Sviluppo Aeronautico Meridionale SpA (Italy)
SW	Short-Wave
SW	Strategic Wing (USAF)
SW	Surface Wave
SWAARM	Smart Weapon Anti-Armour
SWAPO	Southwest Africa People's Organization
SWAT	Special Weapons And Tactics
SWATH	Small Waterplane Area Twin Hull
SWATT	Simulator for Wire-guided Anti-tank Tactical Training
SWB	Short Wheelbase
SWC	Semi-Wadcutter
SWC	Special Weapons Center (USAF)
SWCL	Short Wave Chemical Laser (0.5 to 1.0 micron)
SWCL	Special Warfare Craft, Light (USA)
SWCM	Special Warfare Craft, Medium (USA)
SWD	Stork-Wartsila Diesel BV (Netherlands)
SWD	Surface Wave Device
SWELL	Sea Wolf Enhanced Low-Level Fuze
SWG	Standard Wire Gauge
SWHQ	Static War Headquarters

SWIM	System Wide Integrity Management (USA)
SWIP	Systems Weapons Improvement Programme
SWIPE	Simulated Weapon Impact Predicting Equipment
SWIR	Short-Wave IR
SWL	Strategic Weapon Launcher
SWORD	Small Warship Operational Reconnaissance & Decoy launcher (UK)
SWR	Standing Wave Ratio
SWU	Switching Unit
SX	Sheet Explosive
SXTF	Satellite X-Ray Test Facility
SY	Security branch of INTEL (NATO)
SYLVER	Systeme de Lancement Vertical. Vertical launch system (France)
SyOP	Security Operating Procedure
SYRINX	System Rapide Inter-armees a Base d'engins et Fonctionnant en Bande X. Integrated X-band radar & missile system (France)
SYSCON	System Control Container (Falkland Islands)
SYSTO	Systems Command program officer (USA)
S&A	Safety & Arming
S&TI	Scientific & Technical Intelligence
S+S	Sichelschmidt & Schlasse (Germany)
S/L	Sea Level
S/Lt	Sub Lieutenant
S/Sgt	Staff Sergeant

T

T	Sulphur & Chlorine Compound. Chemical warfare agent
T	Tracer
T2A	Total Terrain Avionics
T45TS	T-45 Training System (USA)
TA	Target Acquisition
TA	Target Alert
TA	Technical Assessment
TA	Telescoped Ammunition
TA	Terrain Avoidance
TA	Territorial Army (UK)
TA	Towed Array
TA	Traffic Advisory
TAA	Total Army Analysis 1996-2001 (US Army)
TAAM	Tactical / Tomahawk Airfield Attack Missile
TAB	Towed Assault Bridge
TABMS	Tactical Air Battle Management System
TABV	Theatre Airbase Vulnerability
Tac	Tactical
TAC	Tactical Air Command
TACAMO	Take Charge And Move Out (USA)
TACAN	Tactical Aid to Navigation
TACBE	Tactical Beacon Equipment, lightweight
TACBUS	Tactical Asynchronous Control Bus system
TACC	Tactical Air Command Center (USAF)
TACCAR	Time-Average Clutter Coherent Airborne Radar
TACCIMS	Theatre Automated Command & Control Information Management System
TACCO	Tactical Co-ordinator
TACCS	Tactical Air Command & Control Specialist
TACDEW	Tactical Advanced Combat Direction & Electronic Warfare Trainer
TACDS	Threat Adaptive Countermeasures Dispensing System
TACE	Transmission Alarm & Control Equipment
TACELIS	Tactical Automated Communications Emitter Location & Identification System
TACELIS	Tactical Emitter Location & Identification System
TACEVAL	Tactical Evaluation, air weapon systems (USAF)
TACFAX	Tactical Digital Facsimile
TACFIRE	Tactical Fire direction system
TACG	Tactical Air Control Group (USAF)
TACINTEL	Tactical Intelligence
TACJAM	Tactical Jammer
TACJS	Tactical Automatic Communications Jamming System
TACMS	Tactical Missile System (USA)
TACNAVMOD	Tactical Navigation Modification
TACNET	Tactical Area Communications Network
TACOM	Tactical Area Communications (Germany)
TACOM	Tactical Command
TACOM	Tank Automotive Command (US Army)

TACOM-EWS	Tactical Communications Electronic Warfare System (USA)
TACOMS	Tactical Communications Systems for the Land Combat Zone
TACON	Tactical Control
TACOR	Threat-Assessment & Control Receiver
TACOS	Tactical Airborne Countermeasures Or Strike
TACP	Theatre Air Control Party
TACPRO	Tactical Procedures Performance Analysis
TACR	Tactical Reconnaissance
TAC-RRS	Tactical Remote Receiving System
TACS	Tactical Air Control Squadron
TACS	Tactical Air Control System (USA)
TACSATCOM	Tactical Satellite Communications (USA)
TACSI	TACS Improvements
TACSIM	Tactical Simulation
TACTAS	Tactical Towed Array Sonar
TACTASS	Tactical Towed Acoustic Sensor System
TACTEC	Totally Advanced Communications Technology (USA)
TACTICOS	Tactical Information & Command System
TACTIFS	Tactical Integrated Flight System
TACTS	Tactical Aircrew Combat Training System
TAD	Tactical Air Defence
TAD	Target Acquisition & Designation
TAD	Target Assembly Data
TAD	Trailing Arm Drive
TADDS	Target Alert Display Data Set (USA)
TADGE	Taiwan Air Defence Ground Environment. Also known as Tien Wuang (Skynet)
TADIL	Tactical Digital Information Link (USA)
TADIXS	Tactical Data Information Exchange System
Tadjet	Transport, Airdrop, Jettison
TADOC	Transportable Air Defence Operations Centre
TADS	Tactical Air Defense Sight (USA)
TADS	Target Acquisition & Designation System (USA)
TADS/PNVS	Target Acquisition & Designator Set / Pilot Night Vision Sensor
TAERS	Tactical Aircrew Eye Respiratory System
TAF	Tactical Air Force
TAFICS	Turkish Air Force Integrated Communications System
TAFIIS	TAF Integrated Information System
TAFSEG	Tactical Air Force Systems Engineering Group
TAG	Tactical Airlift Group
TAG	Target Adaptive Guidance circuit
TAG	Telescoped-Ammunition Gun
TAG	Towed Acoustic Generator
TAI	TUSAS Aerospace Industries Inc. (Turkey)
TAIMS	Three-Axis Inertial Measurement System
TAINS	TERCOM-Aided Inertial Navigation System
TAIRCW	Tactical Air Control Wing
TAIS	Tactical Air Intelligence System
TAIS	Technology Application Information System (USA)
TAIS	Thermal Active Intervention System

TAIT	Tornado Air Interception Trainer
T-AKX	Commercial Roll-on Roll-off ship (USA)
TALAFIT	Tank Level Aiming & Firing Trainer
TALC	Tactical Airlift Center (USAF)
TALCM	Tactical Air-Launched Cruise Missile
TALD	Tactical Airborne Laser Designator
TALD	Tactical Air-Launched Decoy
TALISSI	Tactical Light Shot Simulator
TALT	Tactical Arms Limitations Talks
TAM	Tanque Argentino Mediano. Medium tank (Argentina)
TAM	Tovarna Avtomobilov in Motorjev (Slovenia)
TAM	Towed Acoustic Monitor
TAMP	Terminally-guided Anti-armour Mortar Projectile (UK)
TAMS	Tank Anti-Missile System
TANS	Tactical Air Navigation System
TANZAM	Tanzania – Zambia Railroad Line
TAOC	Tactical Air Operations Centre (USA)
TAOM	Tactical Air Operations Module
TAOR	Tactical Area Of Responsibility
TAP	Toxicological Agents Protective
TAPS	Tracking And Positioning System
TAR	Threat-Avoidance Receiver
TARADCOM	Tank-Automotive Research And Development Command
TARAN	Tactical Radar And Navigation
TARC	Tactical Air Reconnaissance Center (USAF)
TARCAP	Target Combat Aircraft Practice
TARE	Telegraph Automatic Radio Equipment (NATO)
TAREWS	Tactical Air Reconnaissance & Electronic Warfare Support (USAF)
TARG	Telescoped Ammunition Revolver Gun
TARIF	Telegraph Automatic Routeing In the Field (UK)
TARMOS	Tactical Radio Monitoring System
TARPS	Tactical Air Radar Signal Processor
TARPS	Tactical Aircraft Reconnaissance Pod System (US Navy)
TARS	Tactical Air Reconnaissance System
TAS	Tactical Acoustic System
TAS	Target Acquisition System
TAS	Technology Applications & Services Co. (USA)
TAS	Towed Array Sonar (Germany)
TAS	Tracking Adjunct System
TAS	Training Access Switch (NATO)
TAS	Training Agressor Squadron
TAS	True Airspeed. EAS corrected for density appropriate to altitude
TASC	Touch-Activated Simulator Control
TASD	Trajectory And Signature Data
TASES	Tactical Airborne Signal Exploitation System
TASI	True Airspeed Indicator
TASIG	Target And Simulation Image Generator
TASM	Tactical Air-to-Surface Missile
TASM	Tomahawk Anti-Ship Missile (USA)
TASMO	Tactical Air Support for Maritime Operations

TASS	Tactical Air Support Squadron (USAF)
TASS	Tactical Signal Simulator
TASS	Towed Array Surveillance System
TASUMA	Target And Surveillance Un-Manned Aircraft
TASVAL	Tactical Aircraft Survivability Against Armour programme (USA)
TASWIT	Tactical Anti-Submarine Warfare Interim Trainer
TAT	Tactical Armament Turret (USA)
TATS	Tactical Aircraft Training System
TAU	Target Acquisition & tracking Unit
TAU	Terminal Access Unit
TAV	Transatmospheric Vehicle
TAVITAC	Traitement Automatique et Visualisation Tactique (France)
TAWC	Tactical Air Warfare Center
TAWDS	Target Acquisition / Weapon Delivery System
TB	Torpedo-Bomber
TBA	Thomson Brandt Armements (France)
TBA	Tres Bas Altitude. Very low level / altitude (France)
TBAT	TOW / Bushmaster Armored Turret (USA)
TBC	Tactical Bombing Competition
TBC	Tailored-Bloom Chaff
TBC	Toss-Bombing Computer
TBCP	Tele-Brief Control Panel
TBD	Time / Bearing Display
TBD	To Be Decided
TBD	Torpedo Boat Destroyer
TBM	Tactical Ballistic Missile
TBMEWS	Tactical Ballistic Missile Early Warning System
TBPA	Torso Back Protective Armour
TBR	Torpedo-Bomber Reconnaissance
TBSR	Team Battle Shooting Range (UK)
TCA	Time of Closest Approach
TCAC	Tactical Control & Analysis Center (USA)
TCAS	Traffic Alert & Collision Avoidance System
TCB	Target Connection Box
TCB	Trusted Computer Base
TCB	Turret Control Box
TCC	Tactical Co-ordination Console
TCC	Technical & Commercial Consultants SA (Spain)
TCC	Telecommunications Centre
TCC	Transport Control Centre
TCCCS/IRIS	Tactical Command, Control & Communications System / Integrated Radio Intercommunication System (Canada)
TCCF	Tactical Communications Control Facility (USA)
TCCP	Take Command Control Panel
TCD	Landing ship, dock
TCI	Technology for Communications International (USA)
TCI	Turbo-Charged & Intercooled
TCJ	Tactical Communications Jamming
TCO	Tactical Control Officer

TCP	Tactical Computer Processor
TCP	Thermite Case Penetrator
TCP	Travure Courte Portee. Assault treadway bridge (France)
TCP	Turret Control Panel
TCPS	Transportable Collective Protection System
TCP/IP	Transmission Control Protocol / Internet Protocol
TCS	Tactical Computer System (USA)
TCS	Tactical Control Squadron
TCS	Target Control Set
TCS	Television Camera Set
TCS	Tone Commander Systems Inc. (USA)
TCS	Trusted Computer System
TCSS	Torpedo Control System, Submarines (UK)
TCT	Tactical Computer Terminal (USA)
TCT	Tank Crew Trainer
TCT	Target Centred Tracker
TCTI	Transport Canada Training Institute
TCU	Tactical Communications System
TCU	Tactical Control Unit
TCU	Thermal Cueing Unit
TCU	Tracking Control Unit
TCU	Tracking & Communication Unit (UAV)
TC&E	Training Control & Evaluation
TD	Tank Destroyer
TD	Territorial Defense (Slovenia)
TDA	Tactical Decision Aid. Ocean environment
TDA	Theatre Defence Architecture
TDAR	Tactical Defence Alerting Radar
TDBM	Track Database Manager
TDC	Tactical Display Console
TDC	Target Designator Control
TDC	Through-Deck Cruiser
TDCC	Trousse de Detection Chimique de Controle (France)
TDCS	Tank Driver Command System
TDD	Target Detection Device
TDE	Target Data Extractor. Digital interface
TDECC	Tactical Display & Engagement Control Console
TDF	Tactical Digital Facsimile
TDI	Time Delay & Integration
TDLPS	Tactical Data Link Processing System
TDLS	Tactical Delta Loop System
TDM	Tactical Munition Dispenser
TDM	Technical Development & Marketing Inc. (USA)
TDM	Time Division Multiplex
TDMA	Time Division Multiple Access
TDMS	Tactical Data Management System
TDOA	Time Delay Of Arrival
TDP	Target Data Panel
TDR	Target Data Receiver
TDRS	Tracking & Data Relay Satellite system

TDS	Tactical Data System
TDS	Tactical Drone Squadron
TDS	Target Designation Sight
TDT	Tactical Data Terminal
TDT	Tank Driver Trainer
TDU	Terminal Display Unit
TDU	Turret Displacement Unit
TDV	Technology Development Vehicle
TDZ	Total Danger Zone
TD&E	Tactics Development & Evaluation
TE	Tactical Evaluation
TE	Tangent Elevation
TEAMCO	Trans European Airways Maintenance Co. (Belgium)
TECMUS	Tactical Electronic Countermeasures Upgrade System
TECOM	Test & Evaluation Command (US Army)
TED	Tactical / Threat Evaluation Display
TED	Threat Environment Description
TEDA	Triethylenediamine
TEDS	Tactical Expendable Drone System
TEF	Total Environment Facility
TEG	Thermo-Electric Generator
TEI	Trompeter Electronics Inc. (USA)
TEK	Thomson-CSF Elektronik GmbH (Germany)
TEL	Transporter Erector Launcher
TELAR	Transporter, Erector, Launcher And Radar
TELP	Tunnel Entrance for Litter Patients
TEMPER	Tent Extendable Modular Personnel
TEMS	Turbine Engine Monitoring System
TENCAP	Tactical Exploitation of National Capabilities programme (US Army)
TEORS	Tactical Electro-Optical Reconnaissance System
TEOSS	Tactical Emitter Operational Support System (USAF)
TEP	Tactical Electronic Plot
TER	Terrain following radar
TER	Triple Ejector Rack
TERA	Terminal Effects Research & Analysis
TERCOM	Terrain Contour Matching navigation system
TEREC	Tactical Electronic Reconnaissance
TERMM	Transportable Emergency Response Monitoring Module
TERPROM	Terrain Profile Matching
TES	Tactical Environment Simulation
TES	Target Engagement System
TES	Test & Evaluation Squadron
TES	Transportable Earth Station (UK)
TESAC	Training & Evaluation System for Active Countermeasures
TESCO	Toshiba Electronic Systems Co. Ltd (Japan)
TESON	Tropas Especiales de Selva y Operaciones Nocturnas (Honduras)
TET	Turbine Entry Temperature
TETCO	Technical Equipments Trading Co. (Egypt)
TEU	Turret Electronics Unit
TEWA	Threat Evaluation & Weapon Assignment

TEWS	Tactical Electronic Warning System
TEWT	Tactical Exercise Without Troops (UK)
TEXS	Tactical Explosive System
TF	Task Force
TF	Terrain-Following
TFA	Toxic Free Area
TFC	Tactical Fire-Control
TFC	Tactical Fusion Centre (AAFCE)
TFC	Torpedo Fire Control
TFCC	Tactical Flag Command Center
TFCS	Tank Fire Control System
TFD	Terrain Following Device
TFD	Time / Frequency Display
TFE	Terrain-Following E-scope
TFEC	Tactical Fighter Electronic Combat
TFEL	Thin Film Electro-Luminescent
TFF	The Future Frigate
TFG	Tactical Fighter Group
TFM	Tactical Flight Management system
TFMS	Tactical Frequency Management System
TFOCA	Tactical Fibre Optic Cable Assembly
TFOV	Total Field-Of-View
TFPA	Torso Front Protective Armour
TFR	Tank Ferry Raft
TFR	Terrain-Following Radar
TFS	Tactical Fighter Squadron
TFS	Trainer Fighter Simulator
TFSI	Thomson Field Services International (France)
TFT	Thin Film Transistor
TFTS	Tactical Fighter Training Squadron
TFU	Turret FLIR Unit
TFW	Tactical Fighter Wing
TFWC	Tactical Fighter Weapons Center
TF/TA2	Terrain-Following / Terrain Avoidance / Threat Avoidance
TG	Task Group
TGC	Trunk Group Cluster / Communications
TGM	Trunk Group Multiplexer
TGMTS	Tank Gunnery & Missile Target System
TGP	Terminally Guided Projectile
TGS	Tank Gun Sight
TGS	Teplovaya Golvka Samonavedaniya. Single-channel passive infra-red transparent seeker unit (Romania)
TGS	Turkish General Staff
TGSM	Terminally Guided Submunitions
TGT	Target
TGT	Turbine Gas Temperature
Tgt Opp	Target of Opportunity
TGTS	Tank Gunnery Training Simulator
TGU	Torpedo Guidance Unit

TGW	Terminally Guided Weapon
THAAD	Theatre High Altitude Air Defence
THAI	Thai Airways International Ltd
THAWS	Tactical Homing And Warning System
THI	Tactical Hit Indicator
THK	Turk Hava Kuvvetleri
THL	Tourelle Helicoptere Leger. Light helicopter turret (France)
THV	Tres Haute Vitesse. Ammunition (France)
TI	Target Indicator
TI	Thermal Imager (or Imaging)
TI	Thermal Infra-red
TIA	Textile Industries Australia Ltd
TIALD	Thermal Imaging Airborne Laser Designator
TIARA	Tornado Integrated Avionics Research Aircraft (UK)
TIAS	Target Identification & Acquisition System
TIC	Target Insertion Controller
TIC	Tel-Instrument Electronics Corp. (USA)
TICM	Thermal Imaging Common Module
TID	Tactical / Target Information Display
TID	Technical Interface Description
TIDP	Telemetry & Image Data Processing
TIES	Tactical Information Exchange System
TIG	Tesco International Ltd (Saudi Arabia)
TIGER	Terrifically Insensitive to Ground Effect Radar
TIGS	Thermal Imaging Gun Sight
TIJ	Telescope Injection Unit for optical sights
TILAS	Tank Integrated Laser Sight
TILOS	Tangram Integrated Logistic System
TILS	Tactical ILS
TIM	Target Information Module
TIM	Transmission Interface Module
TIMS	Technology Integration of Missile Subsystem
TINA	Thermal Imaging Navigation Aid
TINS	Thermal Imaging Navigation Set
TIP	Technical Improvement Programme
TIP	Tracking & Impact Prediction
TIPI	Tactical Information Processing & Interpretation system (USAF)
TIR	Target Illuminating Radar
TIR	Thermal Imaging Radar
TIR	Thermal Infra-Red
TIR	Tracking & Illuminating Radar
TIRE	Tank Infra-Red Elbow
TIRSS	Theatre Intelligence, Reconnaissance & Surveillance Study (USAF)
TIS	Tactical Input Segment (JSIPS)
TIS	Tactical Intelligence Squadron
TIS	Television Installation Services (Mansfield) Ltd (UK)
TIS	Thermal Imaging System
TIS	Tracking Information / Instrumentation Subsystem
TIS	Trusted Information Systems Inc. (USA)
TISEO	Target Identification System, Electro-Optical

TISH	Thermal Imaging Sensor Head
TISS	Thermal Imaging Security System
TISS	Thermal Imaging Surveillance System
TITE	TEWS Intermediate Test Equipment
TIVS	Thermal Imager Vehicle Sight
TIVS	Thermal Intervention Venting System
TJS	Tactical Jamming System
TLA	Towed Linear-Array sonar
TLAM	Tomahawk Land Attack Missile (USA)
TLAM-C	Tomahawk Land Attack Missile, Conventional (USA)
TLAM-D	Tomahawk Land Attack Missile, sub-munitions warhead (USA)
TLAM-N	Tomahawk Land Attack Missile, Nuclear (USA)
TLB	Trailer-Launched Bridge
TLC	Through Life Costs
TLC	Tonnes, Lifting Capacity
TLC	Top Line Co. (USA)
TLE	Treaty-Limited Equipment
TLFS	Tank Laser Firing Simulator
TLLF	Tactical Low-Level Flying
TLP	Tactical Leadership Programme (AAFCE)
TLR	Tank Laser Rangefinder
TLS	Tactical Landing System
TLS	Tank Laser Sight
TLS	Through Life Support
TLVS	Taktisches Lutverteidigungssytem. Missile system (Germany)
TM	Transverse Magnetic
TMA	Target Motion Analysis
TMB	Techniques Michel Brochier SA (France)
TMC	The Technical Materiel Corp. (USA)
TMD	Tactical Modular Display
TMD	Tactical Munitions Dispenser
TMD	Theatre Missile Defence
TME	Trade Mark Electronics B.V.B.A. (Belgium)
TMEPS	Transverse Mounted Engine Propulsion System
TMLD	Low-level Reporting Service (Germany)
TMLZ	Low-level Reporting Centre (Germany)
TMMI	Trojan Manufacturing & Marketing Inc (Philippines)
TMP	Tactical Machine Pistol
TMP	Telemetrical Measuring Projectile
TMPS	Theater Mission Planning System (USA)
TMS	Training Management System
TMS	Turret Modernisation System
TMU	Transducer Matching Unit. Sonar element
TMX	Tracking Module (Switzerland)
TN	Telenorma GmbH (Germany)
TNF	Tactical (or Theatre) Nuclear Forces
TNFS3	Theatre Nuclear Forces, Survivability, Security & Safety
TNI	Trusted Network Interpretation
TNIU	Trusted Network Interface Unit
TNR	Transfer of control message, Non-Radar

TNSW	Thyssen Nordseewerke GmbH (Germany)
TNT	Trinitrotoluene
TNW	Tactical Nuclear Warfare
TNW	Theatre Nuclear Weapon
T-O	Take-Off
TO	Technical Order (USAF)
TO	Teritorijalna Obramba. Slovenian Civil Defence (Yugoslavia)
TOA	Time Of Arrival
TOAD	Towed Offboard Active Decoy
TOC	Tactical Operations Center (USA)
TOCC	Tactical Operations Control Centre
TOD	Thermal Observation Device
TODS	Tactical Optical Disk System
TOE	Table of Organisation & Equipment
TOF	Time of Flight
TOGS	Thermal Observation & Gunnery System
TOI	Tourelleau D'Observation et D'Intervention. Observation & intervention Cupola (France)
TOJ	Track On Jamming
TOM	Tactical Optronic Mast
TOM	Training Outdoor Military
TOO	Target Of Opportunity
TOP	Total Obscuring Power
TOPAS	Transporter Obrneny Pasovy (Poland)
TOPAT	Towed Passive Torpedo Target
TOPMS	Take-Off Performance Monitoring System
TORT	Tactical Operational Readiness Trainer
TOS	Tactical Operations System (USA)
TOS	Text & Office Systems
TOSS	Television Optical Scoring System
TOT	Time On Target
TOTE	Tracker, Optical Thermally Enhanced
TOTS	Tower Operator Training System
TOW	Tube-launched Optically-tracked Wire-guided missile (USA)
TOYOCOM	Toyo Communication Equipment Co. Ltd (Japan)
TP	Target Practice
TP	Technical Publication
TP	Training Practice
TPA	Target Practice Ammunition
TPAR	Tactical Penetration-Aid Rocket
TPC	Tri-lateral Program office, COBRA
TPCS	Team Portable COMINT System (USMC)
TPCSDS	Target Practice, Cone-Stabilised, Discarding Sabot
TPDF	Tanzanian People's Defence Forces
TPE	Thermoplastic elastomer
TPE	Tracking & Pointing Experiment
TPER	Target Practice, Extended Range
TP-FL	Target Practice, Flash
TPFSDS-T	Target Practice, Fin-Stabilised, Discarding Sabot, Tracer
TPGID	Tank Precision Gunnery Inbore Device

TPH	Tech Pubs Hal Inc. (USA)
TPH	Technical Publications Hal. Inc. (USA)
TPI	Transco Products Inc. (USA)
TPM	Terrain Profile Matching
TPPX	Target Practice, Proximity-fuzed
TPR	Terrain Profile Recorder
TPS	Tank Periscope Sight
TPS	Thermal Protection System
TP-SM	Target Practice, Smoke
TP-SP	Target Practice, Spotting
TP-T	Target Practice, Tracer
TP-T	Training Practice, Tracer
TPT	Target Projectile Training round
TPT	Troop Proficiency Trainer
TPU	Timer Power Unit (components of IED)
TQM	Technical Quartermaster (UK)
TQM	Total Quality Management
TQMS	Three Quays Marine Services Ltd (UK)
TR	Tactical Reconnaissance (USA)
TR	Theatre Reserve
TRA	Technical Research Associates Inc. (USA)
TRAAMS	Time Reference Angle of Arrival Measurement System
TRACA	Total Radar Aperture-Control Antenna
TRACALS	Traffic-Control Approach & Landing System (USAF)
TRACER	Tactical Reconnaissance Armoured Combat Equipment
TRACKSTAR	Tracked Search & Target Acquisition Radar System (USA)
TRACS	Terminal Radar And Control Systems (Canada)
TRACS	Transportable Radar And Communications Simulator
TRADESMAN	Transportable Rapid Deployment Steerable Mount Antenna
TRADOC	Training & Doctrine Command (US Army)
TRAM	Target Recognition Attack Multi-sensor
TRAP	Terminal Radiation Airborne measurements Program
TRC	Transitions Research Corp. (USA)
TRDI	Technical Research & Development Institute (Japan)
TRE	Target-Rich Environment
TREDS	Tactical Reconnaissance Exploitation Demonstration System
TREE	Transient Radiation Effects on Electronics
TREP	Thrusted Replica Decoy
TRES	Tactical Radar & ESM System
TRF	Tactical Replay Facility
TRF	Threat Radar Frequency
TRF	Tuned Radio Frequency
TRIA	Tracking Range Instrumented Aircraft (USAF)
TRIAD	Triple Air Defence
TRIADS	Technique for Reading an Integrated Air Defence System
TRIGAT	Tri-lateral Anti-tank Guided Weapon
TRI-MNC	Tri-Major NATO Commanders
TRI-TAC	Tri-Service Tactical Communications programme (USA)
TRLV	Tracked Rapier Launch Vehicle
TRM	Toutes Roues Motrices (France)

TRN	Terrain Referenced Navigation
TROL	Tape & Rotorless On-Line machine
TROSCOM	Troop Support Command (US Army)
TROSS	Technical & Rear Operations Support System
TROUND	Triangular Round
TRP	Threat Recognition Processor (US Navy)
TRS	Tactical Reconnaissance Squadron
TRS	Tactical Reconnaissance System / Sensor
TRS	Training Reception System
TRSB	Time Reference Scanning Beam
TRSV	Tracked Rapier Support Vehicle
TRT	TEREC Remote Terminal
TRTG	Tactical Radar Threat Generator
TRU	Transmitter / Receiver Unit
TRUD	Time Remaining Until Dive
TRUMP	Tribal Class Upgrade & Modernisation Program (Canada)
TRV	Terminal Rendezvous Point
TS	Transattack Survivability
TS	Transmitter Segment
TS	Transport Squadron
TsAGI	Central Aero-Hydrodynamics Institute (Russia)
TSC	Triple Store Carrier
TSCM	Technical Surveillance Countermeasures
TSCPU	Tactical Sweep Coverage & Planning Unit
TSD	Tactical Situation Display
TSF	Technical Supply Flight (Royal Air Force)
TSFC	Thrust Specific Fuel Consumption
TSFCS	Tank Simplified Fire Control System
TSG	Target Set Group (radar uplink / downlink)
TSG	Tri-Service Group
TSGCE	Tri-Service Group on Communications & Electronics
TSGT	Transportable Satellite Ground Terminal
TSIR	Total System Integration Responsibility (USA)
TSMT FES	Trident Sonar Maintenance Trainer Front End Simulator
TSMTR	Transmitter
TSO	Technical Standard Order (FAA)
TSOR	Tentative Specific Operational Requirement
TSP	Time & Space Processing Inc. (USA)
TSP	Turret Stabilised Platform
TSPR	Total System Performance Responsibility (USA)
TSQ	Time & Superquick
TSR	Tactical Strike / Reconnaissance
TSR	Torpedo Spotter Reconnaissance
TSS	Tactical Surveillance Sonobuoy
TSS	Tethered Satellite System
TSSAM	Tri-Service Stand-off Attack Missile (USA)
TSSR	Tropo-Satellite Support Radio (USA)
TSSTS	Tactical SIGINT System Training Simulator
TSU	Telebriefing Switching Unit
TSU	Telescopic Sight Unit

TSV	Through Sight Video
TSW	Tactical Supply Wing (Royal Air Force)
TT	Team Terminal. Single-channel communications terminal (USA)
TT	Tech-Tool Plastics (USA)
TT	Threat Transmitter
TT	Torpedo Tube
TT	Tous Temps. All-weather (France)
TT	Transport de Troupes. Troop transporter (France)
TTB	Tank Test-Bed (USA)
TTB	Tanker / Transport / Bomber
TTB	Target Triggered Burst
TTBT	Threshold Test Ban Treaty
TTBTS	TTB Training System
TTC	Tactical Training Centre
TTC	Tape Transport Cartridge
TTC	Technical Training Command (Royal Air Force)
TTC	Tracking, Telemetry & Command
TTCS	Target Tracking & Control Station
TTD	Thomson-TRT Defense (France)
TTF	Tanker Task Force
TTF	Target-Towing Flight
TTF	Threat Training Facility (USA)
TTFTC	Tyuratam Flight Test Centre (CIS)
TTG	Time To Go
TTH	Tactical Transport Helicopter
TTL	Target Technology Ltd (UK)
TTL	Transistor / Transistor Logic
TTOMT	Tank Turret Organisational Maintenance Trainer
TTP	The Thompson Partnership (UK)
TTP	Time To Protection
TTR	Target Tracking Radar
TTRD	Transmitter & Tape Recorder Detector
TTS	Tank Thermal Sight
TTS	Time To Station
TTS	Total Training System
TTS	Tube Temperature Sensor
TTT	Teletype Trainer
TTTE	Tri-National Tornado Training Establishment (UK)
TTTS	Tanker Transport Training System
TTW	Transition To War
TTWS	Terminal Threat Warning System
TTY	Teletypewriter
TU	Towed Unit
TUA	TOW Under Armor (USA)
TUAAM	Tuner Unit Automatic Antenna Matching
TUH	Truck, Universal, Heavy (UK)
TUL	Truck, Utility, Light (UK)
TUM	Truck, Utility, Medium (UK)
TUR	Tieffliegler-Uberwachungs-Radar. Low-level surveillance radar (Germany)

TURBO	Thermal Unit & Rangefinder for Battery Observers
TURMS	Tank Universal Reconfigurable Modular System
TUSLOG	The US Logistics Group (USAFE)
TUSSF	Trident Unique Software Support Facility
Tutts	Bowater Tutt Industries Pty Ltd (Australia)
TV	Television
TV	Terminal Velocity
TVA	Target Vector Analysis
TVAT	Television Air Trainer
TVC	Thrust Vector Control
TVCO	Television Converter
TVDS	Tactical Video Distribution System
TVDU	Television Display Unit
T-VIT	Tactical Video Imaging Terminal
TVLA	Tuned Vertical Line Array sonobuoy
TVM	Track-Via-Missile
TVR	Track-Via-missile Radar guidance
TVR	Trajectory / Velocity Radar
TVRS	Tactical Video Receiving System
TVSU	Television Sight Unit
TVT	Television Trainer
TWDS	Tactical Water Distribution System
TWE	Technical Writing & Engineering Ltd (Israel)
TWGSS	Tank Weapon Gunnery Simulation System
TWI	Threat Warning Indicator
TWMP	Track Width Mine Plough
TWOATAF	Second Allied Tactical Air Force (NATO)
TWR	Threat Warning Receiver
TWS	Tail Warning Set
TWS	Threat Warning System
TWS	Track-While-Scan (radar)
TWS	Thermal Weapon Sight
TWSS	TOW Weapon Subsystem
TWT	Travelling Wave Tube
TWTA	Travelling Wave Tube Amplifier
TWU	Tactical Weapons Unit (Royal Air Force)
TWVMP	Tactical Wheeled Vehicle Modernisation Plan
TWX	Theatre War Exercise
TW/AA	Tactical Warning & Attack Assessment (NORAD)
TX	Transmit(ter) / Transmission
TZM	Transportno-Zaryazhayushchaya Mashina. Transport-loader vehicle (CIS)
T/MDV	Towing / Mine Detection Vehicle

U

UAH	Up After Hit
UAI	United Alloys Inc. (USA)
UAMCE	Union Africaine et Malgache de Co-operation Economique
UAMED	Union Africaine et Malgache de banques pour le Development
UAR	Unattended Radar
UARS	Unattended Radar Station
UARS	Unmanned Air Reconnaissance System
UAS	University Air Squadron (UK)
UASTAS	Unmanned Airborne Surveillance & Target Acquisition System (USA)
UAV	Unmanned Air Vehicle
UBE	Ultra Bypass Engine
UCAR	Union of Central African Republics
UCNI	Unified Communications / Navigation / Identification
UCPS	Unhardened Collective Protection System
UCPS(DU)	Unhardened Collective Protection System (for Desert Use)
UCQBR	Urban Close Quarter Battle Range
UCS	Underwater Combat System
UDA	Ulster Defence Association (UK)
UDF	Unducted Fan
UDI	Undersecretariat for Defence Industries (Turkey)
UDL	Unclassified Data Link
UDMH	Unsymmetrical dimethyl hydrazine
UDMU	Universal Decoder Memory Unit
UDR	Ulster Defence Regiment
UDT	Undersea Defence Technology Conference & Exhibition
UEJ	Unattached Expendable Jammer
UEP	Underwater Electric Potential
UET	Universal Engineer Tractor (USA)
UFC	United Fastener Co. Inc. (USA)
UFCP	Up-Front Control Panel
UFDR	Universal Flight Data Recorder
UFF	Ulster Freedom Fighters (UK)
UFH	Ultra-lightweight Field Howitzer
UFN	Until Further Notice
UFO	UHF Follow-On communications satellite programme
UFO	Unidentified Flying Object
UGM	Submarine-launched surface attack missile (USA)
UGS	Upgraded Silo
UHPT	Undergraduate Helicopter Pilot Training
UI	Unicorn International Pte Ltd (Singapore)
UIA	Union of International Associations
UII	United Industries International (Korea, South)
UIL	User Interface Language
UIS	Upper Information System
UIT	Union International de Tire. International shooting standard
UK	United Kingdom of Great Britain & Northern Ireland
UK AIR CCIS	UK Air Command & Control Information System
UKA	Universzalis Kumulativ Akna. Anti-tank mine (Hungary)

UKADGE	UK Air Defence Ground Environment
UKADR	UK NATO Air Defence Region
UKAEA	United Kingdom Atomic Energy Authority
UKAF	United Kingdom Air Force (NATO)
UKAIR	United Kingdom Air Forces CCIS
UKLF	United Kingdom Land Forces
UKLF AMIS	UK Land Forces Automated Management Information System
UKMACCS	UK Maritime Coastal Communication System
UKMCCIS	UK Marines Command, Control & Information System
UKMF	United Kingdom Mobile Force
UKMILREP HQ NATO	UK Military Representative, HQ NATO
UKMSCS	UK Military Satellite Communications System
UKNMR SHAPE	UK National Military Representative, SHAPE
UKRAOC	UK Regional Air Operations Centre
UKSUBCAMS	UK Submarine Communications System
UKWMO	UK Warning & Monitoring Organisation
UK/NLAF	UK / Netherlands Amphibious Force
ULAIDS	Universal Locator Airborne Integrated Data System
ULC	Unit Load Container
ULCS	Unit-Level Circuit Switch
ULEA	Ultra-Long-Endurance Aircraft
ULEV	Unmanned Long-Endurance Vehicle
ULLA	Ultra-Low-Level Airdrop
ULMS	Undersea-Launch / Long-range Missile System
ULMS	Unit-Level Message Switch
ULSA	Ultra-Low Sidelobe Antenna
ULV	Upper Limit of Video
UMA	Unmanned Aircraft
UMA	Unusual Military Activities
UMCP	Unit Maintenance Collection Point
UMD	Unrefuelled Mission Distance
UMIDS	Universal Mine-Dispensing System
UMK	Upravlyaemoye Minnoye Kompleks. Seismic remote control device (CIS)
UML	Union Monetaria Latina
UMOA	Union Monetaire Ouest-Africaine
UMT	Universal Military Training
UN	United Nations
UNA	Ukraine National Assembly
UNAVEM	UN Angola Verification Mission
UNC	United Nations Command
UNCD	United Nations Centre for Disarmament
UNCTAD	United Nations Conference on Trade & Development
UNDOF	United Nations Disengagement Observer Force
UNDP	United Nations Development Programme
UNDRO	United Nations Disaster Relief Office
UNEF	United Nations Emergency Force
UNEPTA	United Nations Expanded Programme of Technical Assistance
UNESCO	United Nations Educational, Scientific & Cultural Organisation
UNFICYP	United Nations Peace-keeping Force in Cyprus

UNFO	Undergraduate Naval Flight Officer (USA)
UNICOM	Unit Computing for the Army (UK)
UNIDO	United Nations Industrial Development Organisation
UNIFIL	United Nations Interim Force in Lebanon
UNIHEDD	Universal Head-Down Display
UNIKOM	UN Iraq / Kuwait Observation Mission
UNIPOM	United Nations India – Pakistan Observation Mission
UNISET	Unified Tactical Communication System (Israel)
UNISIST	United Nations International System of Information on Science & Technology
UNITA	Uniao Nacional para a Independencia Total de Angola
UNITAR	United International Anti-submarine Warfare exercise
UNITAR	United Nations Institute for Training & Research
UNITAS	United International Anti-Submarine warfare exercises
UNMOGIP	United Nations Military Observer Group in India & Pakistan
UNPROFOR	UN Protection Force
UNSCEAR	United Nations Scientific Committee on the Effects of Atomic Radiation
UNSCOB	United Nations Special Committee on the Balkans
UNSDRI	United Nations Social Defence Research Institute
UNSO	Ukrainian People's Self Defence Force
UNT	Undergraduate Navigator Training (USAF)
UNTS	Undergraduate Navigator Training System
UOES	User Operational Evaluation System
UOR	Urgent Operational Requirement
UP	Unguided Projectile
UPCo	Universal Propulsion Co. Inc. (USA)
UPM	Universal Processor Module
UPS	Uniterruptable Power Supply
UPT	Undergraduate Pilot Training
UQ	Ultra Quick fuze
URANCO	British, Dutch, West German Consortium for Processing Uranium
URG	Underway Replenishment Group
UROVESA	Vehiculos Especiales Uro SA (Spain)
URR	Ultra Reliable Radar
URS	User Requirements Specification
URV	Unmanned Research Vehicle
US	United States
USA	United States Army
USA	United States of America
USA	Useful Screen Area
USAAAVS	US Army Agency for Aviation Safety
USAADS	US Army Air Defense School
USAAVSCOM	US Army Aviation Systems Command
USABMD	US Army Ballistic Missile Defense agency
USADC	US Army Air Defense Command
USAEC	US Army Electronics Command
USAF	United States Air Force
USAFA	US Air Force Academy
USAFE	US Air Force Europe

USAFR	US Air Force Reserves
USAFSS	US Air Force Security Service
USAISC	US Army Information Systems Command
USAMC	US Army Missile Command
USAMICOM	US Army Missile Command
USAMRDC	US Army Medical Research & Development Command
USANCA	US Army Nuclear & Chemical Agency
USANG	US Air National Guard
USAR	US Army Reserve
USAREUR	US Army Europe
USARSPACE	US Army Space Command
USASAC	US Army Security Assistance Command
USASDC	US Army Strategic Defense Command
USASTC	US Army Signal Training Centre
USB	Upper SideBand
USBL	Ultrashort Baseline. Torpedo tracking range
USC	United Somali Congress
USCENTCOM	US Central Command
USCG	US Coast Guard
USCGR	US Coast Guard Reserve
USDCH	Universal Self-Deployable Cargo Handler
USEUCOM	US European Command
USF	United Somalia Front
USFJ/5AF	US Forces Japan & 5th Air Force
USGW	Underwater-to-Surface Guided Weapon
USIA	US Information Agency
USIMM	US International Management & Marketing Inc. (USA)
USLO	US Liaison Officer
USLO (SACLANT)	US Liaison Officer, Supreme Allied Commander Atlantic (NATO)
USMA	US Military Academy
USMC	United States Marine Corps
USMCR	USMC Reserves
USMS	Utility Systems Management System
USN	United States Navy
USNAVEUR	US Naval Forces Europe
USNR	US Navy Reserves
USNTPS	US Navy Test Pilot School
USP	United Somali Party
USS	United States Ship
USSP	Universal Sensor Signal Processor
USSPACECOM	US Space Command
USSTAF	US Strategic Air Force
USTO	Ultra-Short Take-Off
USTRANSCOM	US Transportation Command
USTS	UHF Satellite Tracking System
US/VTOL	Ultra-Short or VTOL
UT	Ultrasonic
UT	Universal Time
UTAAS	Universal Tank & Anti-Aircraft System
UTC	United Technologies Corp. (USA)

UTFCS	Universal Tank Fire Control System
UTIO	United Technologies International Operations Inc. (France)
UTM	Universal Transverse Mercator
UTR	Underwater recovery Transponder/Release unit
UTRC	United Technologies Research Center (USA)
UTS	Universal Turret System
UU	Ulster Unionist (UK)
UUO	Union of Ukraine Officers
UUOD	Union of Ukraine Officers in the Draspora
UV	Ultra-Violet
UV	Unmanned Vehicle
UVAS	Unmanned Vehicle for Aerial Surveillance
UVF	Ulster Volunteer Force (UK)
UVL	Ultra-Violet Laser
UVT	MSI-Ulvertech Ltd (UK)
UW	Unconventional Warfare
UWB	Ultra-Wideband
UWS	Upgunned Weapons System. One-man turret
UWW	Underwater Weapons
UXB	Unexploded Bomb

V

VAAC	Vectored-thrust Aircraft Advanced Control
VAB	Vehicule de l'Avant Blinde. Front armoured car (France)
VAB	Vickers Armoured Bridgelayer (UK)
VAD	Vehicule d'Appui Direct (France)
VAD	Velocity / Azimuth Display
VADAR	Vehicule Autonome de Defense Anti-aerienne Rapproche. Autonomous close-action anti-aircraft vehicle (France)
VADM	Vice Admiral (US Navy)
VADS	Vulcan Air Defense System (USA)
VAE	Vehiculo Armado Exploration
VAES	Voice-Activated Electronic System
VAI	Voice-interactive Avionics
VAM	Vehicle Alarm Module
VAMP	VHSIC Avionics Modular Processor
VAMPIR	Veille Air Mer Panoramique Infra-Rouge. Shipborne infra-red surveillance system (France)
VAPE	Vehiculo Apoyo y Exploration
VAR	VHF Aural Range
VARRV	Vickers Armoured Repair & Recovery Vehicle (UK)
VARV	Vickers Armoured Recovery Vehicle (UK)
VAS	Visual Augmentation System
VAS	Voice-Activated System
VAST	Versatile Avionics Shop Tester
VAT	Vehicule Atelier (France)
VATAS	Voice And Tone Annunciation System
VATLS	Visual Airborne Target-Locator System
VATS	Video Augmented Tracking System
VAWS	Voice Alarm Warning System
VBC	Vehicule Blinde de Combat. Armoured combat vehicle (France)
VBL	Vehicule Blinde Leger. Light armoured vehicle (France)
VBM	Volatile Bulk Memory
VBW	Vertical Ballistic Weapon
VCA	Vehicule Chenille d'Accompagnement. Tracked support vehicle (France)
VCASS	Visual Coupled Airborne Systems Simulator
VCC	Veicolo Corazzato de Combattimento (Italy)
VCC	Voice Communication Control system
VCDS	Vice Chief of the Defence Staff (UK)
VCDS	Voice Communications Distribution System
VCFS	Visually Coupled Flight System
VCG	Vehicule de Combat du Genie. Armoured engineer vehicle (France)
VCI	Vehicule de Combat d'Infanterie. Infantry combat vehicle (France)
VCID	Voice-Controlled Interactive Device
VCM	Visual Countermeasures
VCO	Voltage Controlled Oscillator
VCOS	Vice Chief Of Staff
VCP	Vehicle Control Point
VCP	Visual Control Position

VCPC Vehiculo de Combate Puesto de Comando. Command post vehicle (Argentina)
VCR Variable Compression Ratio
VCR Vehicule de Combat a Roues. Wheeled combat vehicle (France)
VCR Video Cassette Recorder
VCR/AT Vehicule de Combat a Roues / Atelier Technique (France)
VCR/IS Vehicule de Combat a Roues / Intervention Sanitaire (France)
VCR/PC Vehicule de Combat a Roues / Poste de Commandement (France)
VCR/TH Vehicule de Combat a Roues / Tourelle HOT (France)
VCR/TT Vehicule de Combat a Roues / Transport de Troupes (France)
VCS Vehicle Control System
VCS Versatile Console System
VCS Video Camera System
VCS Visual Control Station
VCS Voice Communication System
VCTIS Vehicle Command & Tactical Information System
VCTM Vehiculo de Combate Transporte de Mortero. Armoured mortar carrier (Argentina)
VCTP Vehiculo de Combate Transporte de Personal. Armoured personnel carrier (Argentina)
VCTS Variable Cockpit Training System
VCU Vehicular Communication Unit
VCZ Vehiculos de Combate de Zapadores (Spain)
VC/L Vehicle Container / Launcher (UAV)
VDA Vehicule de Defense Anti-aerienne. Anti-aircraft defence vehicle (France)
VDA Versatile Drone Autopilot
VDAA Vehicule d'Auto-Defense Anti-aerienne (France)
VDC Volunteer Defence Corps
VDD Vehicle Detection Device
VDK Vapour Detection Kit
VDM Viscous Damped Mount
VDP Validation Demonstration Phase
VDS Variable Depth Sonar
VDSL Vickers Defense Systems Ltd (UK)
VDT Virtual Dumb Terminal
VDU Visual Display Unit
VEBAL Vertical Ballistic
VEBLIN Vehiculos Blindados SA (Spain)
VEC Vehiculo de Exploracion de Caballerie. Cavalry scout vehicle (Spain)
VECP Value Engineering Change Proposal
VEDES Vehicle Exhaust Dust Ejection System (USA)
VEMASID Vehicle Magnetic Signature Duplicator
VEMS Versatile Exercise Mine System
VER Vertical Ejector Rack
VERTIC Verification Technology Information Centre (UK)
VERTREP Vertical replenishment
VEX Voice Encryption Accessory
VF Voice Frequency
VFCT Voice Frequency Carrier Telegraph

VFMED	Variable Format Message Entry Device
VFO	Variable Frequency Oscillator
VFR	Visual Flight Rules
VG	Vertical Gyro
VGG	Visual Graphics Generator
VGMU	Vulcan Gunner Monitor Unit (USA)
VHI	Vehicle Heading Indicator
VHIS	Visual Hit Indicator System
VHSIC	Very High Speed Integrated Circuit
VI	Visual Identification mode
VIB	Vehicule d'Intervention du Base (France)
VIC	Visual Instrumentation Corp. (USA)
Vice Adm	Vice Admiral
VID	Virtual Image Display
VID	Visual Identification
VIDS	Vehicle Integrated Defense System programme (USA)
VIEW	Video Interpretation Exploitation Workstation
VIEWS	Vibration Indicator Early Warning System
VIFF	Vectoring In Forward Flight
VIGS	Video Interactive Gunnery System
VIM	Verres Industriels SA (Switzerland)
VINS	Vehicle Integrated Navigator System
VIP	Vehicle Improvement Programme
VIRCS	Vehicle Intercommunication & Radio Control System
VIRSS	Visual & Infra-Red Smoke Screening System
VIS	Vehicular Intercommunications System
VIS	Visual
VISAA	Viseur Anti-Arien. Stabilised anti-aircraft sight (France)
VISMIR	Visible to Medium Infra-Red
VISMOD	Visual Modification
VISSR	Visible IR Spin / Scan Radiometer
VISTA	Very Intelligent Surveillance & Target Acquisition
VISTA	Visual Imagery Simulation Training Aid
VIT	Vision In a Turn
VIU	Vehicle Interface Unit
Vl	Relative velocity of flow on the underside of an aerofoil
VL	Vertical Landing
VLA	Vehicule Leger Aeromobile (France)
VLA	Vertical Launch ASROC
VLAD	Vertical Line Arrange DIFAR
VLAST	Video-based Live Aircraft Stinger Trainer
VLC	Vehicule Leger de Combat. Light armoured car (France)
VLC	Very Low Clearance
VLCC	Very Large Container Carrier
VLCHV	Very Low-Cost Harrassment Vehicle
VLEA	Very-Long-Endurance Aircraft
VLF	Vectored-Lift Fighter
VLGM	Vertical Loading Gun Mount
VLLR	Very Light Laser Rangefinder
VLOC	Very Large Oil Carrier

VLRA	Vehicule Leger de Reconnaissance et d'Appui (France)
VLS	Vertical Launch System (missile)
VLSI	Very Large Scale Integration
VLSMS	Vehicle Launched Scatterable Mines System
VLTT	Vehicule de Liaison Tout Terrain (France)
VM-AT	Versatile Multi-Aimer Trainer
VMBT	Vickers Main Battle Tank (UK)
VMC	Vibration Mountings & Controls Inc. (USA)
VMD	Video Motion Detector
VMDA	Vehicle-Mounted Decontamination Apparatus
VMDI	Vector Miss-Distance Indicator
VME	Virtual Memory Environment
VMI	Voltage Multipliers Inc. (USA)
VMM	Variable Moments Magnet
VMRMDS	Vehicle-Mounted Road Mine Detector System
VMS	Vehicle Management System
VMS	Vehicle Motion Sensor
VMT	Vehicle Manoeuvring Trainer
VMU	Velocity Measuring Unit
VMU	Voice Message Unit
VNAS	Vehicle Navigation Aid System
VNAV	Vertical Navigation
VNL	Vertical Navy Launcher
Vo	Muzzle Velocity
VO	Visual Optics
VOA	Artillery Observation Vehicle (France)
VOA	Velocity Of Arrival
VOC	Visual Optical Countermeasures
VOCODER	Voice encoder / decoder
VOCOM	Voice Communication system
VOD	Vertical On-board Delivery
VOGAD	Voice Operated Gain Adjustment Device
Vol	Volunteer
VOL	Vertical On-board Landing
VOLCAN	Viseur Optronique Leger pour Conduite d'Arme Navale. Naval fire control system (France)
VOR	Vehicle Off Road
VOR	VHF Omnidirectional Range (radio beacon)
VORTAC	VHF Omnidirectional Range Tacan
VOR/LOC	VHF Omnidirectional Range Locator
VOS	Voice Operated Switch
Vox	Voice keying or activation
vp	Variable pitch
VP	Vital Point
VP	Vulnerable Point
VPAF	Vietnamese People's Air Force
VPI	Vertical Position Indicator
VPN	Vietnamese People's Navy
VPU	Voice Projection Unit

VPV	Vyprost'ovaci Pasove Vozidlo. Tracked armoured vehicle (Czechoslovakia)
VPVO	Vojska Protivovozdushnoj Oborony. Air defence troops (CIS)
VR	Voice Recognition
VR	Volunteer Reserve
VRDE	Vehicle Research & Development Establishment
VRECS	Variable Radiated Emission Control System
VRI	Vanguard Research Inc. (USA)
VRL	Vehicule Reconnaissance Leger. Light reconnaissance vehicle (France)
VROC	Vertical Rate Of Climb
VRP	Visual Reporting Posts
VRRTFL	Variable Reach Rough Terrain Forklift
VRRTFLT	Variable Reach Rough Terrain Forklift Truck
VRU	Vertical Reference Unit
VS	Velocity Search
VS	Vestigial Sideband
VSCF	Variable-Speed Constant-Frequency
VSD	Vertical Situation Display
VSD	Video Symbology Display
VSDC	Vehicle Systems Development Corp. (USA)
VSER	Vertical Speed & Energy Rate
VSFI	Vertical-Scale Flight Instrument
VSHORAD	Very Short Range Air Defence
VSI	Vertical Speed Indicator
VSI	Very Seriously Injured
VSLD	Vertical Search Look-Down
VSM	Vehicle Sous-Marine. Underwater vehicle (France)
VSRAD	Very Short-Range Air Defence
VSS	Vehicle Systems Simulator
VSS	Video Shooting System
VSS	Video Signal Simulator
VSSA	Variable-Stability Simulator Aircraft
VST	Variable-Stability Trainer
VSTAR	Variable Search & Track Air defence Radar
VSTT	Variable Speed Training Target
VSW	Verification Software
VSYS	Veda Systems Inc. (USA)
VT	Variable Time
VTA	Voice Terrain Advisory
VTAS	Visual Target Acquisition System
VTL	Vehicle Tactical Logistic
VTL	Vehicule de Transport Logistique. Transport Vehicle (France)
VTM	Voltage Tunable Magnetron
VTMS	Vessel Traffic Management System
VTO	Vertical Take-Off
VTOL	Vertical Take-Off & Landing
VTP	Vehicule Transport de Troupe. Personnel carrier (France)
VTR	Video Tape Recorder
VTRPS	Video Target Range Projection System

VTS	Video Target Simulator
VTT	Vehicule Transport de Troupe. Troop transporter (France)
VTT	Video Target Trainer
V-VS	Voyeonno – Vozdushnyye Sily. Air Force (CIS)
VVSS	Vertical Volute Spring Suspension
VWB	Voice Warning Box
VWC	Vulcan Wheeled Carrier (USA)
VWG	Vereinigung Wehrtechnisches Gerat (Germany)
VWS	Air-Ventilated Wet Suit
VX	Chemical warfare nerve agent (V-agent)
VXB	Vehicule Blinde a Vocations Multiples. Multi-purpose armoured car (France)
V/H	Velocity / Height (ratio)
V/STOL	Vertical / Short Take-Off & Landing

W

W	Weapon
W	Wing
WAA	Wide Aperture Array
WAA	World Aviation Associates (Netherlands)
WAAM	Wide Area Anti-Armour Munitions
WAAS	Wide-Area Active Surveillance
WADIBUOY	Wave Directional Buoy
WADS	Wideband Audio / Data Switch
WAF	Women in the Air Force (USA)
WAFS	Women's Auxiliary Ferrying Squadron (USA)
WAIN	Wide-Area Integrated Network
WAM	Wide Area Mine (USA)
WAMS	Weapon-Aiming Mode Selector
WAN	Wide-Area Network
WAPC	Wheeled Armoured Personnel Carrier
WARF	Wide Aperture Research Facility
WARHUD	Wide-Angle Raster HUD
WARMAPS	Wartime Manpower Planning System (USA)
WARPAC	Warsaw Pact (now defunct)
WAS	Weapon-Aiming System
WASP	War Air Service Program (USA)
WASP	Weasel Attack Signal Processor
WASP	Wide Angle Stinger Pointer
WASP	Wide-Area Special Projectile (USAF)
WASPM	Wide Area Side Penetrating Mine
WAS/MTI	Wide Area Surveillance / Moving Target Indicator
WBC	Weight & Balance Computer
WBK	Wehrbereichskommando (Germany)
WBS	Weight & Balance System
WBSS	Wideband digital Switching System
WBSV	Wideband Secure Voice
WC	Wadcutter. Ammunition
WCCS	Wireless Communication & Control System
WCG	Water-Cooled Garment
WCM	Weapon Control Module
WCMS	Weapons Control & Management System
WCNS	Weapon Control & Navigation System
WCP	Water Carriage Pack
WCS	Waveguide Communications System
WCS	Weapon Control System
WD	War Department
WD	Warning Display
WDAU	Weapon-Dispenser Arming Device
WDC	Weapon Delivery Computer
WDE	Weapon Direction Equipment, naval (USA)
WDIP	Weapon Data Input Panel
WDNS	Weapon Delivery & Navigation System
WDS	Weapon Direction System, naval (USA)

WDS	Weapon Display Subsystem
WE	Weald Electronics Ltd (UK)
WEAL	Westinghouse Electric Australasia Ltd
WECMC	Wing Electronic-Combat Management Course
WEDS	Weapons Effect Display System
WEFT	Wings, Engines, Fuselage, Tail
WEM	Warning Electronic Module
WEP	War Emergency Power
WEP	Weapon Effect Planning
WES	Waterways Experimental Station (USA)
WES	Weapons Effect Simulation
WESIM	Westinghouse Electronic Systems International Marketing Co. (USA)
WESS	Weapons Effects Signature Simulator
WESTLANT	Western Atlantic Area (NATO)
WEU	Warning Electronic Unit
WEU	Western European Union
WEWO	Wing Electronic Warfare Officer (Royal Air Force)
WFC	World Food Council
WFCS	Warning & Fire Control Station
WFOV	Wide Field-Of-View
WFP	World Food Program
WFSV	Wheeled Fire Support Vehicle
WG	Wandel & Goltermann Inc. (USA)
WG	Working Group
Wg Cdr	Wing Commander
WGC	Wyman-Gordon Co. (USA)
WGD	Windshield Guidance Display
WHAT	Wheels, Hull, Armour, Turret
WHCA	White House Communications Agency (USA)
WHDE	Weapon Handling & Discharge Equipment
WHIDDS	War Headquarters Information Display & Dissemination System (NATO)
WHN	Wolf Hirth GmbH (Germany)
WHO	World Health Organisation
WHQ	War Headquarters
WI	Warning Installations
WI	Wetzel International Inc. (USA)
WIA	Wounded In Action
WIDE	Wide-angle Infinity Display Equipment
WIFT	Westland Indirect Fire Trainer
WIG	Wing In Ground Effect
WIMS	Worldwide Intratheatre Mobility Study
WIN	WWMCCS Intercomputer Network
WINDII	Wind Imaging Interferometer
WINGS	Warning & Intelligence Gathering Sensor programme (US Navy)
WINTEX	Winter Exercise
WIPO	World Intellectual Property Organisation
WIS	Wireless Intercommunications System
WIS	WWMCCS Information System (USA)
WIS JPMO	WIS, Joint Program Management Office (USA)

WIU	Weapon Interface Unit
WJAC	Women's Junior Air Corps (UK)
WL	Water Line
WLDP	Warning Light Display Panel
WLR	Weapon Locating Radar
WMD	Weapons of Mass Destruction
WMO	World Meteorological Organisation
WMRV	Wheeled Maintenance & Recovery Vehicle
WMS	Water Management System
WNC	Weapons & Navigation Computer
WO	Warrant Officer
WO1	Warrant Officer 1st Class
WO2	Warrant Officer 2nd Class
WOC	Wing Operations Center (USA)
WOD	Wind Over Deck
WOMBLE	Wire Operated Mobile Bomb Lifting Equipment
WORD	Wind-Orientated Rocket Deployment
WORM	Write Once, Read Multiple
WOT	Wide-Open Throttle
WP	Warsaw Pact (now defunct)
WP	White Phosphorus
WP	Working Party
WPC	World Peace Council
WPI	Wavell Processor Installation
WPM	Words Per Minute
Wpn	Weapon
WPR	Wideband Programmable Receiver
WP-T	White Phosphorus Tracer
WPU	Weapon Programming Unit
WQAU-P	Water Quality Analysis Unit – Purification
WR	Wilcoxon Research (USA)
WRA	Weapon Replaceable Assembly
WRAC	Women's Royal Army Corp (UK)
WRAF	Women's Royal Air Force (UK)
WRCS	Weapon Release Computer Set (USAF)
WRD	War Reserve Drop-tank
WRE	Weapons Research Establishment (Australia)
WRM	War Reserve Material
WRNS	Women's Royal Naval Service (UK)
WRS	War Reserve Stock
WRSK	War-Readiness Spares / Supply Kit (USAF)
WRVS	Women's Royal Voluntary Service (UK)
WS	Weapon System
WSAL	Westland System Assessment Ltd (UK)
WSC	Weapon System Controller
WSCE	Weapons System Control Element
WSCP	Warning System Control Panel
WSD	World Sales Development (USA)
WSDA	Weapon System Data Acquisition unit
WSDL	Weapon Systems Data Link

WSEG	Weapon Systems Evaluation Group (USA)
WSEP	Weapon System Evaluation Programme
WSF	Weapon Storage Flight (Royal Air Force)
WSIAO	Weapon Support, Improvements & Analysis Office (USA)
WSIP	Weapons System Improvement Programme
WSL	Weapon System Level
WSLI	Warning System on Laser Illumination (Romania)
WSMC	Western Space & Missile Center (USAF)
WSME	Wire Sweep Monitoring Equipment
WSMIS	Weapon System Management Information System
WSMR	White Sands Missile Range (US Army)
WSO	Weapon System Operator / Officer
WSPS	Weapon System Physics Section
WSPS	Wire-Strike Protection System
WSSD	Weapon System Support Development
WSSG	Warning System Signal Generator
WSSP	Weapon System Support Programme
WSSS	Weapon Storage & Security System
WST	Weapon System Trainer
WSTI	Welcom Software Technology International (UK)
WSTR	Weapon System Training Rig
WTD 61	Wehrtechnische Dienststelle fur Luftfahrzeuge (Germany)
WTI	Weapon & Tactics Instructor
WTT	World Trade Transport Corp. (USA)
WVR	Within Visual Range
WWABNCP	Worldwide Airborne National Command Post (USA)
WWACPS	Worldwide Airborne Command Post System
WWMCCS	Worldwide Military Command & Control System
WWMCS	Worldwide Military Communication System
WWO	Wing Weapon Officer (Royal Air Force)
WWS	Wild Weasel Squadron (USAF)
WWSVA	Worldwide Secure Voice Architecture (USA)
WX	Airborne Weather Radar
W/Cdr	Wing Commander

X

X	Experimental
XAC	Xi'an Aircraft Co. (China)
XAGM	Experimental Air-to-Ground Missile
XAIM	Experimental Air Intercept Missile
XBT	Expendable Bathythermograph
XCTD	Expendable Conductivity, Temperature & Depth profiling system
XMGM	Experimental Mobile-launched surface attack Guided Missile
XO	Executive Office
XRL	X-Ray Laser
XSS	Expendable Sonobuoy / Sonar / Sonic Sensor
XSTAT	Expendable Submarine Tactical Transceiver
XSV	Expendable Sound Velocity system

YAG	Miscellaneous auxiliary craft (USA)
YAG	Yttrium-Aluminium-Garnet
YC	Open lighter (USA)
YCF	Car Float (USA)
YCV	Aircraft transportation lighter (USA)
YD	Floating crane (USA)
YDT	Diving Tender (USA)
YF	Covered lighter (USA)
YFB	Ferry Boat or launch (USA)
YFD	Yard Floating Dock (USA)
YFN	Covered lighter (USA)
YFNB	Large covered lighter (USA)
YFND	Dry dock companion craft (USA)
YFNX	Lighter, special purpose (USA)
YFP	Floating Power Barge (USA)
YFR	Refrigerated covered lighter (USA)
YFRN	Refrigerated covered lighter (USA)
YFRT	Covered lighter, range tender (USA)
YFU	Harbour utility craft (USA)
YG	Garbage lighter (USA)
YGN	Garbage lighter (USA)
YHLC	Salvage Lift Craft, Heavy (USA)
YIG	Yttrium Indium Garnet
YM	Dredge (USA)
YMLC	Salvage Lift Craft, Medium (USA)
YMRS	Yugoslav Multiple Rocket System
YNG	Gate craft (USA)
YO	Fuel oil barge (USA)
YOG	Gasoline barge (USA)
YOGN	Gasoline barge (USA)
YON	Fuel oil barge (USA)
YOS	Oil Storage barge (USA)
YP	Patrol craft (USA)
YP	Yield Point
YPA	Yugoslav People's Army
YPD	Floating Pile Driver (USA)
YR	Floating workshop (USA)
YRB	Repair & Berthing barge (USA)
YRBM	Repair, Berthing & Messing barge (USA)
YRDH	Floating dry dock workshop, hull (USA)
YRDM	Floating dry dock workshop, machine (USA)
YRR	Radiological Repair barge (USA)
YRST	Salvage craft tender (USA)
YS	Yield Strength
YSD	Seaplane wrecking derrick (USA)
YSR	Sludge Removal barge (USA)
YTB	Large harbour tug (USA)
YTL	Small harbour tug (USA)

YTM	Medium harbour tug (USA)
YW	Water barge (USA)
YWN	Water barge (USA)

Z

ZAB	Zazhigatel'naya Aviabomba. Aerial incendiary bomb (CIS)
ZA-PVO	Zenitnaya Artilleriya-PVO. Anti-aircraft artillery (CIS)
ZAR	Zeus Acquisition Radar
ZD	Zero Defects
ZELL	Zero-Length Launching
ZELMAL	Zero-Length Launch & Mat Landing
ZETA	Zero Energy Thermonuclear Apparatus
ZF	Zahnradfabrik Friedrichshafen (Germany)
ZF	Zero Force
ZFW	Zero-Fuel Weight
ZI	Zone of the Interior (USA)
ZNA	Zimbabwe National Army
ZOA	Zone Of Alarm
ZODIAC	Zone Digital Automatic Communications network (Netherlands)
ZP	Zone of Protection
ZPI	Zone Position Indicator radar
ZRB	Zenitnaya Raketnaya Brigada. Air defence missile brigade (CIS)
ZRD-SD	Zenitnaya Raketnye Kompleks – Strednoye Deiestive. Anti-aircraft missile system – medium range (CIS)
ZRV	Zenitnaya Raketnye Voiska. Zenith Rocket Troops (CIS)
ZRV-PVO	Zenitnaya Raketnye Voiska-PVO. Anti-aircraft missile troops (CIS)
ZST	Zone Standard Time
ZSU	Zenitnaya Samokhodnaia Ustanovka (CIS)
ZULU	Greenwich Meantime
ZVEI	Zentralverband der Elektrotechnik – und Elektronikindustrie e.V. (Germany)

APPENDIX 1

OFFICER RANKS FOR NATO COUNTRIES

BELGIUM

ARMY	NAVY	AIR FORCE
Luitenant-Generaal	Vice-Admiraal	Luitenant-Generaal
Generaal-Majoor	Divisie-Admiraal	Generaal-Majoor
Brigade-Generaal	Commodore	Brigade-Generaal
Kolonel	Kapitein-ter-zee	Kolonel
Luitenant-Kolonel	Fregat-Kapitein	Luitenant-Kolonel
Majoor	Korvette-Kapitein	Majoor
Kapitein-Commandant	Luitenant-ter-zee 1ste klas	Kapitein-Commandant
Kapitein	Luitenant-ter-zee	Kapitein
Luitenant	Vaandrig-ter-zee	Luitenant
Onder-Luitenant	Vaandrig-ter-zee 2de klas	Onder-Luitenant

CANADA

ARMY	NAVY	AIR FORCE
General	Admiral	General
Lieutenant-General	Vice-Admiral	Lieutenant-General
Major-General	Rear-Admiral	Major-General
Brigadier-General	Commodore	Brigadier-General
Colonel	Captain	Colonel
Lieutenant-Colonel	Commander	Lieutenant-Colonel
Major	Lieutenant-Commander	Major
Captain	Lieutenant	Captain
Lieutenant	Sub-Lieutenant	Lieutenant
Second-Lieutenant	Acting-Sub-Lieutenant	Second-Lieutenant

DENMARK

ARMY	NAVY	AIR FORCE
General	Admiral	General
Generallojtnant	Viceadmiral	Generallojtnant
Generalmajor	Kontreadmiral	Generalmajor
Brigadegeneral	Kommandor	Oberst
Oberst	Kommandorkaptajn	Oberstlojtnant
Oberstlojtnant	Orlogskaptajn	Major
Major	Kaptajnlojtnant	Kaptajn
Kaptajn	Premierlojtnant	Premierlojtnant
Premierlojtnant	Lojtnant	Lojtnant
Lojtnant	Sekondlojtnant	Sekondlojtnant
Sekondlojtnant		

FRANCE

ARMY

General d'Armee
General de Corps
General de Division
General de Brigade
Colonel
Lieutenant-Colonel
Chef de Bataillon
Capitaine
Lieutenant
Sous-Lieutenant

NAVY

Amiral
Vice-Amiral d-Escadre
Vice-Amiral
Contre-Amiral
Capitaine de Vaisseau
Capitaine de Fregate
Capitaine de Corvette
Lieutenant de Vaisseau
Enseigne de Vaisseau 1ere Classe
Enseigne de Vaisseau 2eme Classe

AIR FORCE

General d'Armee Aerienne
General de Corps Aerien
General de Division Aerienne
General de Brigade Aerienne
Colonel
Lieutenant-Colonel
Commandant
Capitaine
Lieutenant
Sous-lieutenant

GERMANY

ARMY

General
Generalleutnant
Generalmajor
Brigadegeneral
Oberst
Oberstleutnant
Major
Hauptmann
Oberleutnant
Leutnant

NAVY

Admiral
Vizeadmiral
Konteradmiral
Flottilenadmiral
Kapitan zur See
Fregattenkapitan
Korvettenkapitan
Kapitanleutnant
Oberleutnant zur See
Leutnant zur See

AIR FORCE

General
Generalleutnant
Generalmajor
Brigadegeneral
Oberst
Oberstleutnant
Major
Hauptmann
Oberleutnant
Leutnant

GREECE

ARMY

Stratigos
Antistratigos
Ypostratigos
Taxiarchios
Syntagmatarchis
Antisyntagmatarchis
Tagmatarchis
Lochagos
Ypolochagos
Anthypolochagos

NAVY

Navarchos
Antinavarchos
Yponavarchos
Archipliarchos
Pliarchos
Antipliarchos
Plotarchis
Ypopliarchos
Anthypopliarchos
Anthyposminagos

AIR FORCE

Pterarchos
Antipterarchos
Ypopterarchos
Taxiarchos
Sminarchos
Antisminarchos
Espisminagos
Sminagos
Yposminagos

ITALY

ARMY

Generale di Corpo d'Armata
Generale di Divisione
Generale di Brigata
Colonnello Comandante di Corpo
Colonnello
Tenente Colonnello Comandante di Corpo
Tenente Colonnello
Maggiore
Primo Capitano
Capitano
Tenente
Sottotenente

NAVY

Ammiraglio di Squadra
Ammiraglio di Divisione
Contrammiraglio
Capitano di Vascello
Capitano di Fregata
Capitano di Corvetta
Tenente di Vascello
Sottotenente di Vascello

AIR FORCE

Generale di Squadra Aerea
Generale di Divisione Aerea
Generale di Brigata Aerea
Colonnello
Tenente Colonnello
Maggiore
Capitano
Tenente
Sottotenente

LUXEMBOURG

ARMY

Commandant de l'Armee
Colonel
Major
Capitaine
Premier Lieutenant
Lieutenant

NETHERLANDS

ARMY

Generaal
Luitenant-generaal
Generaal-majoor
Brigade-generaal
Kolonel
Luitenant-kolonel
Majoor
Kapitein
Kapitein
Eerste Luitenant
Tweede Luitenant

NAVY

Admiraal
Luitenant-admiraal
Vice-admiraal
Schout-bij-nacht
Commandeur
Kapitein ter zee
Kapitein-luitenant ter zee
Luitenant ter zee der eerde klasse
Luitenant ter zee der tweede klasse
Luitenant ter zee der derde klasse

AIR FORCE

Generaal
Luitenant-generaal
Generaal-majoor
Commodore
Kolonel
Luitenant-kolonel
Majoor
Kapitein
Eerste Luitenant
Tweede Luitenant

NORWAY

ARMY

General
Generalloytnant
Generalmajor
Oberst I
Oberst II
Oberstloytnant
Major
Kaptein
Loytnant
Fenrik

NAVY

Admiral
Viseadmiral
Kontreadmiral
Kommandor
Kommandor Kaptein
Orlogskaptein
Kapteinloytnant
Loytnant
Fenrik
Ustskrevet

AIR FORCE

General
Generalloytnant
Generalmajor
Obesrt I
Oberst II
Oberstloytnant
Major
Kaptein
Loytnant
Fenrik

PORTUGAL

ARMY

General
Brigadeiro
Coronel
Tenente-Coronel
Major
Capitao
Tenente

NAVY

Almirante de Armada
Almirante
Vice-Almirante
Contra-Almirante
Capitao-de-Mar-e-Guerra
Capitao-de-Fragata
Capitao-Tenente
Primeiro-Tenente
Segundo-Tenente
Guarda-Marinha

AIR FORCE

Marechal
General
Brigadeiro
Coronel
Tenente-Coronel
Major
Capitao
Tenente
Alferes

SPAIN

ARMY

Capitan General
Teniente General
General de Division
General de Brigada
Coronel
Teniente Coronel
Commandante
Capitan
Teniente
Alferez
Subteniente

NAVY

Capitan General
Almirante
Vice-Almirante
Contra-Almirante
Capitan de Navio
Capitan de Fragata
Capitan de Corbeta
Capitan
Alferez de Navio
Alferez de Fragata

AIR FORCE

Capitan General
Teniente General
General de Division
General de Brigada
Coronel
Teniente Coronel
Commandante
Capitan
Teniente
Alferez

TURKEY

ARMY	NAVY	AIR FORCE
Maresal	Buyuk Amiral	Maresal
Orgeneral	Oramiral	Orgeneral
Korgeneral	Koramiral	Korgeneral
Tumgeneral	Tumamiral	Tumgeneral
Tuggeneral	Tugamiral	Tuggeneral
Albay	Albay	Albay
Yarbay	Yarbay	Yarbay
Binbasi	Binbasi	Binbasi
Yuzbasi	Yuzbasi	Yuzbasi
Ustgemen	Ustegmen	Ustegmen
Tegmen	Tegmen	Tegmen
Astegmen	Astegmen	Astegmen

UNITED KINGDOM

ARMY	NAVY	AIR FORCE
Field Marshal	Admiral of the Fleet	Marshal of the Royal Air Force
General	Admiral	Air Chief Marshal
Lieutenant-General	Vice-Admiral	Air Marshal
Major-General	Rear-Admiral	Air Vice-Marshal
Brigadier	Commodore	Air Commodore
Colonel	Captain	Group Captain
Lieutenant-Colonel	Commander	Wing Commander
Major	Lieutenant-Commander	Squadron Leader
Captain	Lieutenant	Flight Lieutenant
Lieutenant	Sub-Lieutenant	Flying Officer
Second Lieutenant	Midshipman	Pilot Officer

UNITED STATES

ARMY	NAVY	AIR FORCE
General of the Army	Fleet Admiral	General of the Air Force
General	Admiral	General
Lieutenant-General	Vice-Admiral	Lieutenant-General
Major-General	Rear-Admiral	Major-General
Brigadier	Commodore	Brigadier-General
Colonel	Captain	Colonel
Lieutenant-Colonel	Commander	Lieutenant-Colonel
Major	Lieutenant-Commander	Major
Captain	Lieutenant	Captain
First-Lieutenant	Lieutenant, Junior Grade	First-Lieutenant
Second-Lieutenant	Ensign	Second-Lieutenant

APPENDIX 2

JETDS CODING SYSTEM

The Joint Electronics Type Designation System (JTEDS) is an unclassified US military system designed to identify, by a series of letters and numbers, the use of an equipment and its major components.

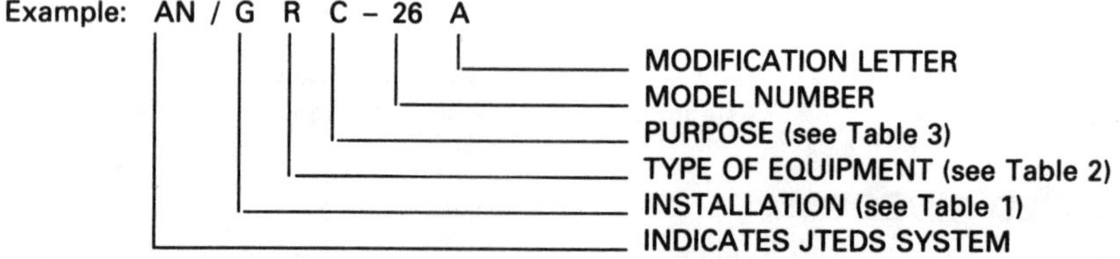

Table 1. INSTALLATION

A	Piloted aircraft
B	Underwater mobile (submarine)
C	Air transportable
D	Pilotless carrier
F	Fixed, ground
G	Ground, general
K	Amphibious
M	Mobile, ground
P	Pack, portable
S	Water surface craft
T	Transportable, ground
U	General utility
V	Vehicular, ground
W	Water (surface & underwater combination)
Z	Piloted & pilotless airborne vehicle combination

Table 2. TYPE OF EQUIPMENT

A	Invisible light, heat radiation
B	Pigeon
C	Carrier
D	Radiac
E	Nupac
F	Photographic
G	Telegraph or teletype
I	Interphone, public address
J	Electromechanical
K	Telemetering
L	Countermeasures
M	Meteorological
N	Sound in air
P	Radar
Q	Sonar, underwater sound
R	Radio
S	Special or combination of types
T	Telephone (wire)
V	Visual & visible light
W	Armament
X	Facsimile or television
Y	Data processing

Table 3. PURPOSE

A	Auxiliary assembly
B	Bombing
C	Communications
D	Direction finding, reconnaissance, surveillance
E	Ejection, release
G	Fire control or searchlight direction
H	Recording, reproducing
K	Computing
L	Searchlight control (inactive)
M	Maintenance, test assembly
N	Navigation aid
P	Reproducing (inactive)
Q	Special or combination of purposes
R	Receiving, passive detection
S	Detecting, range & bearing search
T	Transmitting
W	Automatic flight control, remote control
X	Identification & recognition
Y	Surveillance & control

APPENDIX 3

US MILITARY AIRCRAFT DESIGNATIONS

Designations for US military aircraft conform to this common system.All aircraft are assigned a letter to indicate the basic missionfollowed by a model number. Other letters may then be prefixed toindicate a modified mission or special status. Between the basic missionletter and the type number, a further letter may also be added toindicate a particular vehicle type.

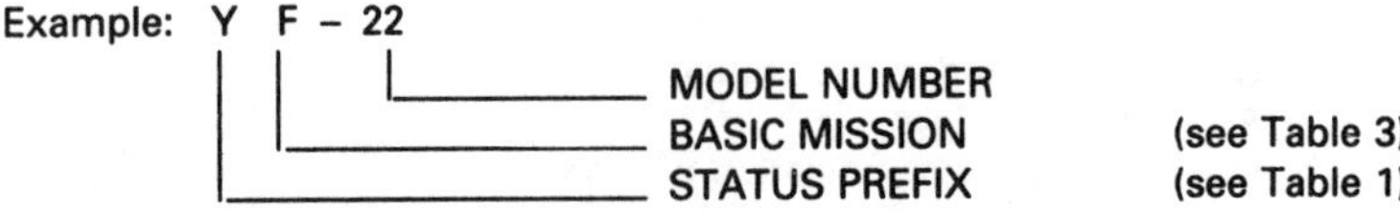

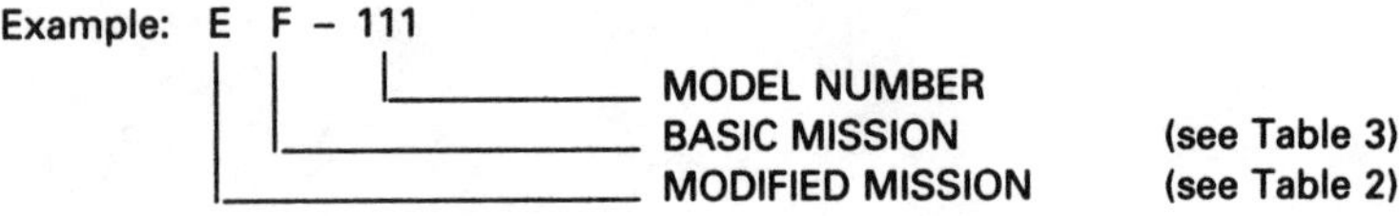

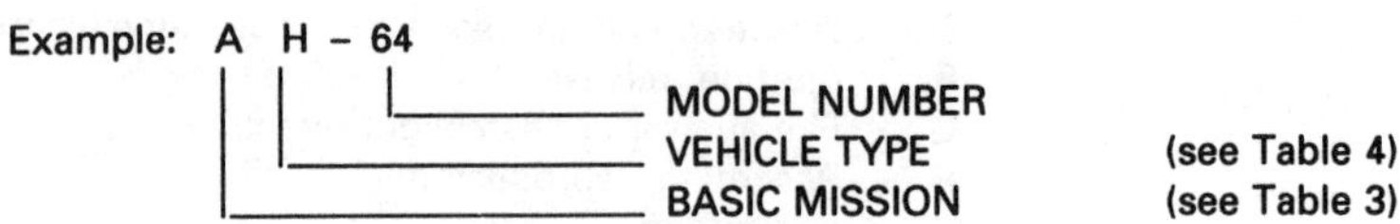

Table 1. STATUS PREFIX

G	Permanently Grounded
J	Special test (temporary)
N	Special test (permanent)
X	Experimental
Y	Prototype
Z	Planning

Table 2. MODIFIED MISSION

A	Attack
C	Transport
D	Director
E	Special electronic installation
F	Fighter
H	Search and rescue
K	Tanker
L	Cold weather
M	Multi-mission
O	Observation
P	Patrol
Q	Drone
R	Reconnaissance
S	Anti-submarine
T	Trainer
U	Utility
W	Weather

Table 3. BASIC MISSION

A	Attack
B	Bomber
C	Transport
E	Special electronic installation
F	Fighter
O	Observation
P	Patrol
R	Reconnaissance
S	Anti-submarine
T	Trainer
U	Utility
X	Research

Table 4. VEHICLE TYPE

G	Glider
H	Helicopter
V	VTOL/STOL
Z	Lighter-than-air vehicle

APPENDIX 4

US MISSILE & RPV DESIGNATIONS

Designations for US missiles and RPVs conform to this common system.A basic designation comprises three letters (indicating launchenvironment, mission and type) followed by a model number. This maybe prefixed with an additional letter to indicate special status.

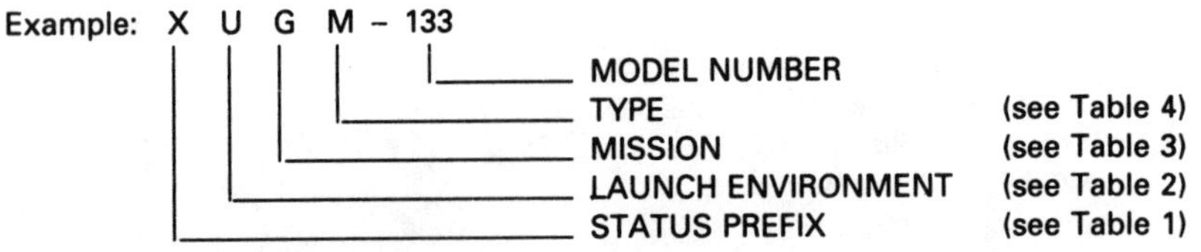

Table 1. STATUS PREFIX

C Captive
D Dummy
J Special test (temporary)
M Maintenance
N Special test (permanent)
X Experimental
Y Prototype
Z Planning

Table 2. LAUNCH ENVIRONMENT

A Air
B Multiple
C Coffin
F Individual
G Runway
H Silo-stored
L Silo-launched
M Mobile
P Soft pad
R Ship
U Underwater attack

Table 3. MISSION

D	Decoy
E	Special electronic installation
G	Surface attack
I	Aerial intercept
Q	Drone
T	Training
U	Underwater attack
W	Weather

Table 4. TYPE

M	Guided missile/drone
N	Probe
R	Rocket

APPENDIX 5

NATO REPORTING NAMES FOR FORMER SOVIET AIRCRAFT

NATO NAME	SOVIET NAME	TYPE
Backfire	Tu-22M	Bomber
Badger	Tu-16	Bomber
Beagle	Il-28/H-5	Bomber
Bear	Tu-95/Tu-142	Bomber
Bison	M-4	Bomber
Blackjack	Tu-160	Bomber
Blinder	Tu-22	Bomber
Brewer	Yak-28	Bomber
Fagot	MiG-15	Fighter
Farmer	MiG-19	Fighter
Fencer	Su-24	Fighter
Fiddler	Tu-28	Fighter
Firebar	Yak-28P	Fighter
Fishbed	MiG-21	Fighter
Fitter A	Su-7	Fighter
Fitter C	Su-17/Su-22	Fighter
Flagon	Su-15	Fighter
Flanker	Su-27	Fighter
Flogger	MiG-23/MiG-27	Fighter
Forger	Yak-38	Fighter
Foxbat	MiG-25	Fighter
Foxhound	MiG-31	Fighter
Fresco	MiG-17	Fighter
Frogfoot	Su-25	Fighter
Fulcrum	MiG-29	Fighter
Cab	Li-2	Transport
Camber	Il-86	Transport
Camel	Tu-104	Transport
Camp	An-8	Transport
Candid	Il-76	Transport
Careless	Tu-154	Transport
Cash	An-28	Transport
Cat	An-10	Transport
Charger	Tu-144	Transport
Clank	An-30	Transport
Classic	Il-62	Transport
Cline	An-32	Transport
Clobber	Yak-42	Transport
Clod	An-14	Transport
Coaler	An-72/An-74	Transport
Cock	An-22	Transport
Codling	Yak-40	Transport

NATO NAME	SOVIET NAME	TYPE
Coke	An-24	Transport
Colt	An-2	Transport
Condor	An-124	Transport
Cookpot	Tu-124	Transport
Coot	Il-18/Il-20	Transport
Crate	Il-14	Transport
Creek	Yak-12	Transport
Crusty	Tu-134	Transport
Cub	An-12	Transport
Curl	An-26	Transport
Halo	Mi-26	Helicopter
Hare	Mi-1	Helicopter
Harke	Mi-10	Helicopter
Havoc	Mi-28	Helicopter
Haze	Mi-14	Helicopter
Helix	Ka-27/Ka-28	Helicopter
Hermit	Mi-34	Helicopter
Hind	Mi-24	Helicopter
Hip	Mi-8	Helicopter
Hokum	Ka-50	Helicopter
Hoodlum	Ka-26	Helicopter
Hook	Mi-6	Helicopter
Hoplite	Mi-2	Helicopter
Hormone	Ka-25	Helicopter
Hound	Mi-4	Helicopter
Madge	Be-6	Maritime
Maestro	Yak-28U	Trainer
Mail	Be-12	Maritime
Mainstay	Il-76	AEW&C
May	Il-38	Maritime
Mascot	Il-28U/HJ-5	Trainer
Maya	L-29	Trainer
Midas	Il-78	Tanker
Midget	MiG-15UTI	Trainer
Mongol	MiG-21H	Trainer
Moose	Yak-11	Trainer
Moss	Tu-126	AEW&C
Moujik	Su-7U	Trainer

APPENDIX 6

NATO REPORTING NAMES FOR FORMER SOVIET MISSILES

NATO NAME	US NAME	SOVIET NAME	TYPE
Acrid	AA-6	R-40	Air-to-air
Alamo	AA-10	AKU-470	Air-to-air
Alkali	AA-1	RS-2U/K-5	Air-to-air
Amos	AA-9		Air-to-air
Anab	AA-3		Air-to-air
Apex	AA-7	R-23/K-23	Air-to-air
Aphid	AA-8	R-60/AK-60	Air-to-air
Archer	AA-11	AKU-72	Air-to-air
Ash	AA-5		Air-to-air
Atoll	AA-2	RS-3/K-13	Air-to-air
Gadfly	SA-11		Surface-to-Air
Gadfly	SA-N-7		Surface-to-Air
Gainful	SA-6	Kub/Kvadrat	Surface-to-Air
Galosh	ABM-1	UR-96	Surface-to-Air
Gammon	SA-5	S-200 Volga	Surface-to-Air
Ganef	SA-4	Krug	Surface-to-Air
Gaskin	SA-9	Strela-1	Surface-to-Air
Gazelle	ABM-3/SH-08		Surface-to-Air
Gecko	SA-8	Romb	Surface-to-Air
Gecko	SA-N-4	Osa-M	Surface-to-Air
Giant	SA-12b		Surface-to-Air
Gladiator	SA-12a		Surface-to-Air
Goa	SA-3	S-125 Neva	Surface-to-Air
Goa	SA-N-1	M-1 Volga	Surface-to-Air
Goblet	SA-N-3		Surface-to-Air
Gopher	SA-13	Strela-10	Surface-to-Air
Grail	SA-7	Strela-2	Surface-to-Air
Grail	SA-N-5	Strela-2M	Surface-to-Air
Gremlin	SA-14	Strela-3	Surface-to-Air
Gremlin	SA-N-8		Surface-to-Air
Grumble	SA-10		Surface-to-Air
Grumble	SA-N-6		Surface-to-Air
Guideline	SA-2	V-75 Dvina	Surface-to-Air
Guideline	SA-N-2	M-2 Dvina	Surface-to-Air
Guild	SA-1	R-113	Surface-to-Air
Kangaroo	AS-3		Air-to-ground
Karen	AS-10		Air-to-ground
Kedge	AS-14		Air-to-ground
Kegler	AS-12		Air-to-ground
Kelt	AS-5		Air-to-ground
Kennel	AS-1		Air-to-ground
Kent	AS-15	RK-55	Air-to-ground

Kerry	AS-7	Grom	Air-to-ground
Kilter	AS-11		Air-to-ground
Kingbolt	AS-13	Kh-59/X-59	Air-to-ground
Kingfish	AS-6		Air-to-ground
Kipper	AS-2		Air-to-ground
Kitchen	AS-4	Burya	Air-to-ground
Kyle	AS-9		Air-to-ground
Saber	SS-20	RSD-10 Pioner	Surface-to-surface
Saddler	SS-7		Surface-to-surface
Sagger	AT-3	9k11/9M14 Malyutka	Anti-tank
Salish	SSC-2a		Surface-to-surface
Samlet	SSC-2b		Surface-to-surface
Sampson	SS-N-21		Surface-to-surface
Sandal	SS-4	R-12	Surface-to-surface
Sandbox	SS-N-12		Surface-to-surface
Sapwood	SS-6		Surface-to-surface
Sark	SS-N-5		Surface-to-surface
Sasin	SS-8		Surface-to-surface
Satan	SS-18	RS-20	Surface-to-surface
Savage	SS-13	RS-12/SS-12	Surface-to-surface
Sawfly	SS-N-8		Surface-to-surface
Saxhorn	AT-7	9K115/9M115 Metis	Anti-tank
Scaleboard A	SS-12a		Surface-to-surface
Scaleboard B	SS-12b	OTR-22	Surface-to-surface
Scalpel	SS-24		Surface-to-surface
Scamp	SS-14		Surface-to-surface
Scapegoat	SS-14		Surface-to-surface
Scarab	SS-21	OTR-21 Tochka	Surface-to-surface
Scarp	SS-9		Surface-to-surface
Scrooge	SS-15		Surface-to-surface
Scrubber	SS-N-1		Surface-to-surface
Scud A	SS-1b	R-11	Surface-to-surface
Scud B	SS-1c	R-17	Surface-to-surface
Scunner	SS-1		Surface-to-surface
Sego	SS-11		Surface-to-surface
Sepal	SSC-1b		Surface-to-surface
Sepal	SS-N-3b	P-35	Surface-to-surface
Serb	SS-N-6		Surface-to-surface
Shaddock	SSC-1a		Surface-to-surface
Shaddock	SS-N-3a	P-6	Surface-to-surface
Shaddock	SS-N-3c	P-7	Surface-to-surface
Shipwreck	SS-N-19		Surface-to-surface
Shyster	SS-3		Surface-to-surface
Sibling	SS-2		Surface-to-surface
Sickle	SS-25	RS-12M	Surface-to-surface
Silex	SS-N-14		Surface-to-surface
Sinner	SS-16	RS-14	Surface-to-surface
Siren	SS-N-9		Surface-to-surface
Skean	SS-5	R-14 Vertikal	Surface-to-surface
Skiff	SS-N-23		Surface-to-surface

Slingshot	SSC-X-4	RK-55	Surface-to-surface
Snapper	AT-1	2K15/2M2 Shmeyl	Anti-tank
Snipe	SS-N-17		Surface-to-surface
Songster	AT-8	Kobra?	Anti-tank
Spandrel	AT-5	9P148/9M113 Konkurs	Anti-tank
Spanker	SS-17	RS-16	Surface-to-surface
Spider	SS-23	OTR-23	Surface-to-surface
Spigot	AT-4	9K111/9M111 Fagot	Anti-tank
Spiral	AT-6	Kohon/Skorpion	Anti-tank
Stallion	SS-N-16		Surface-to-surface
Starbright	SS-N-7		Surface-to-surface
Starfish	SS-N-15		Surface-to-surface
Stiletto	SS-19	RS-18	Surface-to-surface
Stingray	SS-N-18	RSM-50	Surface-to-surface
Sturgeon	SS-N-20	RSM-52	Surface-to-surface
Styx	SSC-3		Surface-to-surface
Styx	SS-N-2a	P-15	Surface-to-surface
Styx	SS-N-2b		Surface-to-surface
Styx	SS-N-2c	P-20/P-21	Surface-to-surface
Sunburn	SS-N-22		Surface-to-surface
Swatter	AT-2	3K11/3M11 Fleyta	Anti-tank

APPENDIX 7

FREQUENCY BANDS & DESIGNATIONS

Current NATO frequency band designations:

A	0 – 250 MHz
B	250 – 500 MHz
C	500 MHz – 1 GHz
D	1 – 2 GHz
E	2 – 3 GHz
F	3 – 4 GHz
G	4 – 6 GHz
H	6 – 8 GHz
I	8 – 10 GHz
J	10 – 20 GHz
K	20 – 40 GHz
L	40 – 60 GHz
M	60 – 100 GHz

Old frequency band designations:

I	100 – 150 MHz
G	150 – 225 MHz
P	225 – 390 MHz
L	390 MHz – 1.55 GHz
S	1.55 – 5.2 GHz
X	5.2 – 10.9 GHz
K	10.9 – 36 GHz
Q	36 – 46 GHz
V	46 – 56 GHz

Other commonly used frequency band designations:

ELF	0 – 3 kHz
VLF	3 – 30 kHz
LF	30 – 300 kHz
MF	300 kHz – 3 MHz
HF	3 – 30 MHz
VHF	30 – 300 MHz
UHF	300 MHz – 3 GHz
SHF	3 – 30 GHz
EHF	30 – 300 GHz

APPENDIX 8

BODY ARMOUR THREAT LEVELS

The following notation is used to indicate the level of effective protection provided by body armours:

Fragment proof	–	Withstands attack by various fragments averaging 1.0 g in weight and moving at an average velocity of 400 m/s; lead shot for small game (but not buckshot) with a maximum velocity of 400 m/s and fired from more than 15 m range.
Level I	–	.22 Long Rifle standard lead bullet; .22 Long Rifle high velocity; .22 Winchester Magnum Full Metal Jacket (FMJ); .38 Special lead bullet; .45 Auto Colt Pistol FMJ.
Level II/IIA	–	.22 Winchester Magnum FMJ; .38 Special FMJ, .38 Special Norma Soft-point; .357 Magnum Jacketed Soft-point (JSP), FMJ and Jacketed Hollow Point (JHP); 9 mm Federal FMJ; 9 mm ATE Federal FMJ; .44 Magnum Norma JSP.
Level III	–	.357 Magnum Norma FMJ; 9 mm Norma FMJ; 9 mm ATE FMJ; .44 Magnum Norma JSP; .44 Magnum Remington JSP.
Level IIIA	–	.357 Magnum Norma FMJ; 9 mm Norma FMJ; 9 mm ATE FMJ; .44 Magnum Norma JSP; .44 Magnum Remington; 12-gauge 00 Buckshot; 12-gauge Brenneke rifled slug.
Level IVA	–	.357 Magnum Metal Piercing (MP); .357 Magnum SFM THV; .38 Special THV; 7.65 auto THV; 9 mm THV; .45 ACP THV.
Level IVB/C	–	7.62 x 39 mm FMJ; .30 US Carbine FMJ; 7.62 x 51 mm NATO FMJ; 12-gauge Blondeau slug; 5.56 x 45 mm NATO FMJ.
Level V	–	7.62 x 39 mm AP; 7.62 x 51 mm AP; .30-06 AP; 5.56 x 45 mm AP.

APPENDIX 9

UNITS OF MEASUREMENT

ABBREVIATION/UNIT		MEASUREMENT
A	ampere	electric current
AC	alternating current	electricity
Ah	ampere hour	electric current
atm	standard atmosphere	pressure
B	bel	sound pressure
Bd	baud	signalling velocity
bhp	brake horsepower	power
bps	bits/second	data processing
BTU	British Thermal Unit	energy
C	celsius	temperature
C	coulomb	electric charge
cal	calorie	energy
cd	candela	illumination
cl	centilitre	volume
cm	centimetre	length
cp	candlepower	illumination
cps	cycles/second	frequency
cwt	hundredweight	weight
daN	decanewton	force
dB	decibel	sound pressure
DC	direct current	electricity
dm	decimetre	length
dP	differential pressure	pressure
dwt	deadweight tonnage	weight
ehp	equivalent horsepower	power
ekW	equivalent kilowatt	power
emf	electromotive force	force
eV	electron volt	energy
F	fahrenheit	temperature
F	farad	electrical capacitance
FL	foot lambert	illumination
fl oz	fluid ounce	volume
ft	foot	length
g	gram, gramme	weight
g	gravity	force
gal	gallon	volume
GHz	gigahertz	frequency
gpm	gallons/minute	volume
grt	gross registered tonnage	weight
h	hour	time
H	henry	inductance
hp	horsepower	power
Hz	hertz	frequency

ABBREVIATION/UNIT		MEASUREMENT
ihp	indicated horsepower	power
in	inch	length
ips	instructions/second	data processing
ISA	Int'l Standard Atmosphere	pressure
J	joule	energy
K	kelvin	temperature
kb	kilobyte	data processing
kg	kilogram	weight
kHz	kilohertz	frequency
kips	thousand instructions/second	data processing
kJ	kilojoule	energy
km	kilometre	length
kmh	kilometres/hour	velocity
kN	kilonewton	force
kops	thousand operations/second	data processing
kph	kilometres/hour	velocity
kt	knot	velocity
kT	kiloton	explosive power
kV	kilovolt	electromotive force
kVA	kilovolt-ampere	electrical power
kW	kilowatt	power
l	litre	volume
lb	pound	weight
lm	lumen	illumination
lx	lux	illumination
m	metre	length
M	mach (speed of sound)	velocity
mb	millibar	pressure
Mbit	megabit	data processing
Mbyte	megabyte	data processing
mg	milligram	weight
MHz	megahertz	frequency
min	minute	time
Mips	million instructions/second	data processing
MJ	megajoule	energy
ml	millilitre	volume
mm	millimetre	length
Mops	million operations/second	data processing
mph	miles/hour	velocity
ms	millisecond	time
mW	milliwatt	power
MW	megawatt	power
N	newton	force
nm	nautical mile	length
ns	nanosecond	time
Ns	newton second	thrust
oz	ounce	weight
P	poise	viscosity
Pa	pascal	pressure

ABBREVIATION/UNIT		MEASUREMENT
pF	picofarad	electrical capacitance
pps	pulses per second	frequency
psi	pounds/square inch	pressure
pt	pint	volume
rad	radian	angle
rpm	revolutions/minute	rotation
rps	revoultions/second	rotation
s	second	time
S	Siemens	conductivity
S	stoke	viscosity
shp	shaft horsepower	power
sr	steradian	angle
st	stone	weight
t	tonne	weight
T	tesla	magnetic flux
tpi	threads/inch	screw thread
t/c	thickness to chord	ratio
u	atomic mass unit	mass
V	volt	electromotive force
VA	volt-ampere	electrical power
vpm	vibrations/minute	vibration
W	watt	power
Wb	weber	magnetic flux
yd	yard	length

PREFIXES (MULTIPLES & SUB-MULTIPLES)

E	exa	million million million
P	peta	thousand million million
T	tera	million million
G	giga	thousand million
M	mega	million
k	kilo	thousand
h	hecto	hundred
da	deca	ten
d	deci	tenth
c	centi	hundredth
m	milli	thousandth
μ	micro	millionth
n	nano	thousand millionth
p	pico	million millionth
f	femto	thousand million millionth
a	atto	million million millionth

APPENDIX 10

CONVERSION FACTORS

WEIGHT

1 Grain	=	0.0648	Gram
1 Gram	=	15.43	Grains
1 Gram	=	0.03527	Ounces
1 Kilogram	=	1000	Grams
1 Kilogram	=	2.205	Pounds
1 Tonne	=	1000	Kilograms
1 Tonne	=	0.9842	Tons (UK)
1 Tonne	=	1.102	Tons (US)
1 Dram	=	1.772	Grams
1 Ounce	=	16	Drams
1 Ounce	=	28.35	Grams
1 Pound	=	453.6	Grams
1 Pound	=	0.4536	Kilograms
1 Pound	=	0.00045	Tonnes
1 Stone	=	14	Pounds
1 Stone	=	6.35	Kilograms
1 Quarter	=	2	Stones
1 Quarter	=	12.7	Kilograms
1 Hundredweight	=	4	Quarters
1 Hundredweight	=	50.8	Kilograms
1 Ton (UK)	=	20	Hundredweight
1 Ton (UK)	=	1.016	Tonnes
1 Ton (US)	=	2000	Pounds
1 Ton (US)	=	907.18	Kilograms

VOLUME

1 Cubic centimetre	=	0.06102	Cubic inches
1 Cubic metre	=	35.32	Cubic feet
1 Cubic metre	=	1.306	Cubic yards
1 Litre	=	61.02	Cubic inches
1 Litre	=	0.22	Imperial gallons
1 Litre	=	0.264	US gallons
1 Pint	=	20	Fluid ounces
1 Pint	=	0.568	Litre
1 Quart	=	2	Pints
1 Quart	=	1.136	Litres
1 Imperial gallon	=	4	Quarts
1 Imperial gallon	=	4.546	Litres
1 Peck	=	2	Gallons
1 Peck	=	9.092	Litres
1 Bushel	=	4	Pecks
1 Bushel	=	36.4	Litres
1 Quarter	=	8	Bushels

VOLUME

1 Quarter	=	291.2	Litres
1 US Gallon	=	3.785	Litres
1 Cubic inch	=	16.39	Cubic centimetres
1 Cubic inch	=	0.01639	Litres
1 Cubic foot	=	0.028	Cubic metres
1 Cubic yard	=	0.7646	Cubic metres
1 Barrel of oil	=	42	US gallons
1 Barrel of oil	=	35	Imperial gallons

LENGTH

1 Millimetre	=	0.03937	Inches
1 Centimetre	=	10	Millimetres
1 Centimetre	=	0.3937	Inches
1 Metre	=	100	Centimetres
1 Metre	=	3.28	Feet
1 Metre	=	1.094	Yards
1 Kilometre	=	1000	Metres
1 Kilometre	=	0.6214	Miles
1 Kilometre	=	0.539	Nautical miles
1 Inch	=	25.4	Millimetres
1 Foot	=	12	Inches
1 Foot	=	0.3048	Metres
1 Yard	=	3	Feet
1 Yard	=	0.9144	Metres
1 Fathom	=	6	Feet
1 Chain	=	22	Yards
1 Furlong	=	10	Chains
1 Mile	=	8	Furlongs
1 Mile	=	1760	Yards
1 Mile	=	1.609	Kilometres
1 Mile	=	0.868	Nautical miles
1 Nautical mile	=	1.852	Kilometres
1 Nautical mile	=	1.152	Miles

ANGLE

1 Radian	=	57.3	Degrees
1 Degree	=	17.45	Milliradians
1 Degree	=	60	Arc minutes
1 Milliradian	=	3.438	Arc minutes
1 Arc minute	=	0.2909	Milliradians
1 Arc minute	=	60	Arc seconds

VELOCITY

1 Radian/second	=	57.3	Degrees/second
1 Degree/second	=	0.01745	Radians/second

ANGLE

1 Centimetre/second	=	0.033	Feet/second
1 Metre/second	=	3.281	Feet/second
1 Kilometre/hour	=	0.621	Miles/hour
1 Kilometre/hour	=	0.539605	Knots
1 Foot/second	=	30.48	Centimetres/second
1 Foot/second	=	0.305	Metres/second
1 Mile/hour	=	1.609	Kilometres/hour
1 Mile/hour	=	0.868	Knots
1 Knot	=	1.852	Kilometres/hour
1 Knot	=	1.152	Miles/hour

AREA

1 Square centimetre	=	0.155	Square inches
1 Square kilometre	=	0.3861	Square miles
1 Square metre	=	10.76	Square feet
1 Square metre	=	1.196	Square yards
1 Hectare	=	10000	Square metres
1 Hectare	=	2.471	Acres
1 Square inch	=	6.452	Square centimetres
1 Square foot	=	0.0929	Square metres
1 Square yard	=	0.8361	Square metres
1 Square mile	=	2.59	Square kilometres
1 Acre	=	4840	Square yards
1 Acre	=	4047	Square metres
1 Acre	=	0.4047	Hectares

POWER, FORCE, ENERGY

1 Kilowatt	=	1.34102	Horespower
1 Horsepower	=	0.7457	Kilowatts
1 Newton	=	0.2248	Foot pounds
1 Foot pound	=	4.448	Newtons
1 Joule	=	0.2388	Calories
1 Kilojoule	=	0.9478	British Thermal Unit
1 British Thermal Unit	=	1.055	Kilojoules
1 Calorie	=	4.1868	Joules

TEMPERATURE

Celsius × 9/5 + 32	=	Fahrenheit
Fahrenheit − 32 x 5/9	=	Celsius

APPENDIX 11

PERIODIC TABLE

SYMBOL	ELEMENT	ATOMIC NUMBER	ATOMIC MASS
Ac	Actinium	89	(227)
Ag	Silver	47	107.868
Al	Aluminium	13	26.9815
Am	Americium	95	(243)
Ar	Argon	18	39.948
As	Arsenic	33	74.9216
At	Astatine	85	210
Au	Gold	79	196.9665
B	Boron	5	10.81
Ba	Barium	56	137.34
Be	Beryllium	4	9.0122
Bi	Bismuth	83	208.9806
Bk	Berkelium	97	(249)
Br	Bromine	35	79.904
C	Carbon	6	12.011
Ca	Calcium	20	40.08
Cd	Cadmium	48	112.4
Ce	Cerium	58	140.12
Cf	Californium	98	(251)
Cl	Chlorine	17	35.453
Cm	Curium	96	(247)
Co	Cobalt	27	58.9332
Cr	Chromium	24	51.996
Cs	Caesium	55	132.9055
Cu	Copper	29	63.546
Dy	Dysprosium	66	162.5
Er	Erbium	68	167.26
Eu	Europium	63	151.96
F	Fluorine	9	18.9984
Fe	Iron	26	55.847
Fm	Fermium	100	(253)
Fr	Francium	87	(223)
Ga	Gallium	31	69.72
Gd	Gadolinium	64	157.25
Ge	Germanium	32	72.59
H	Hydrogen	1	1.008
He	Helium	2	4.0026
Hf	Hafnium	72	178.49
Hg	Mercury	80	200.59
Ho	Holmium	67	164.9303
I	Iodine	53	126.9045
In	Indium	49	114.82
Ir	Iridium	77	192.22

SYMBOL	ELEMENT	ATOMIC NUMBER	ATOMIC MASS
K	Potassium	19	39.102
Kr	Krypton	36	83.8
La	Lanthanum	57	138.9055
Li	Lithium	3	6.941
Lr	Lawrencium	103	(. . .)
Lu	Lutetium	71	174.97
Md	Mendelevium	101	(256)
Mg	Magnesium	12	24.305
Mn	Manganese	25	54.938
Mo	Molybdenum	42	95.94
N	Nitrogen	7	14.0067
Na	Sodium	11	22.9898
Nb	Niobium	41	92.9064
Nd	Neodymium	60	144.24
Ne	Neon	10	20.179
Ni	Nickel	28	58.71
No	Nobelium	102	(254)
Np	Neptunium	93	(237)
O	Oxygen	8	15.9994
Os	Osmium	76	190.2
P	Phosphorus	15	30.9738
Pa	Protactinium	91	231.0359
Pb	Lead	82	207.2
Pd	Palladium	46	106.4
Pm	Promethium	61	(145)
Po	Polonium	84	(210)
Pr	Praseodymium	59	140.9077
Pt	Platinum	78	195.09
Pu	Plutonium	94	(242)
Ra	Radium	88	226.0254
Rb	Rubidium	37	85.4678
Re	Rhenium	75	186.2
Rh	Rhodium	45	102.9055
Rn	Radon	86	(222)
Ru	Ruthenium	44	101.07
S	Sulphur	16	32.06
Sb	Antimony	51	121.75
Sc	Scandium	21	44.9559
Se	Selenium	34	78.96
Si	Silicon	14	28.086
Sm	Samarium	62	150.4
Sn	Tin	50	118.69
Sr	Strontium	38	87.62
Ta	Tantalum	73	180.9479
Tb	Terbium	65	158.9254
Tc	Technetium	43	(99)
Te	Tellurium	52	127.6
Th	Thorium	90	232.0381

SYMBOL	ELEMENT	ATOMIC NUMBER	ATOMIC MASS
Ti	Titanium	22	47.9
Tl	Thallium	81	204.37
Tm	Thulium	69	168.9342
U	Uranium	92	238.029
V	Vanadium	23	50.9414
W	Tungsten	74	183.85
Xe	Xenon	54	131.3
Y	Yttrium	39	88.9059
Yb	Ytterbium	70	173.04
Zn	Zinc	30	65.37
Zr	Zirconium	40	91.22

NOTE: Values in brackets indicate mass numbers of most stable isotopes

APPENDIX 12

INTERNATIONAL PHONETIC ALPHABET

LETTER	WORD	PRONUNCIATION
A	Alfa	AL fah
B	Bravo	BRAH voh
C	Charlie	CHAR lee / SHAR lee
D	Delta	DELL tah
E	Echo	ECK oh
F	Foxtrot	FOKS trot
G	Golf	GOLF
H	Hotel	hoh TELL
I	India	IN dee ah
J	Juliett	JEW lee ETT
K	Kilo	KEY loh
L	Lima	LEE mah
M	Mike	MIKE
N	November	no VEM ber
O	Oscar	OSS cah
P	Papa	pah PAH
Q	Quebec	keh BECK
R	Romeo	ROW me oh
S	Sierra	see AIR rah
T	Tango	TANG go
U	Uniform	YOU nee form / OO nee form
V	Victor	VIK tah
W	Whiskey	WISS key
X	X-ray	ECKS ray
Y	Yankee	YANG key
Z	Zulu	ZOO loo

NOTE: The syllables to be emphasised are shown in capitals

APPENDIX 13

COMPANY TYPES & ABBREVIATIONS

COUNTRY	COMPANY TYPE		OWNERSHIP
Australia	Ltd	Limited	Public
	Pty Ltd	Proprietary Limited	Private
Austria	AG	Aktiengesellschaft	Public
	GmbH	Gesellschaft mit Beschrankter Haftung	Private
Belgium	SA	Societe Anonyme	Public
	NV	Naamloze Vennootschap	Public
	SA/NV		Public
	Sprl	Societe Privee a Responsabilite Limitee	Private
	BVBA	Besloten Vennootschap met Beperkte Aansprakelijkheid	Private
Brazil	SA	Sociedade Anonima	Public
	Ltda	Limitada	Private
	SQRL	Sociedade por Quotas de Responsabilidade Limitada	Partnership
Canada	Ltd	Limited	Public
	Inc.	Incorporated	Private
	Corp.	Corporation	Private
Denmark	A/S	Aktieselskab	Public
	ApS	Anpartsselskab	Private
Finland	Oy	Osakeyhtio	Public
	AB	Aktie Bolag	Public
France	SA	Societe Anonyme	Public/Private
	Sarl	Societe a Responsabilite Limitee	Private
Germany	AG	Aktiengesellschaft	Public/Private
	GmbH	Gesellschaft mit Beschrankter Haftung	Private
	KG	Kommanditgesellschaft	Private
Greece	SA	Societe Anonyme	Public/Private
	SAIC		
	EPE	Etairia Periorismenis fthynis	Private
India	plc	Public Limited Company	Public
	Ltd	Limited	Private
Indonesia	BHD	Berhad	Public
	PT	Perseroan Terbetas	Private

COUNTRY	COMPANY TYPE		OWNERSHIP
Italy	SpA	Societa per Azioni	Public
	Srl	Societa Reponsibilita Limitata	Private
Japan	KK	Kabushiki Kaisha	Public
	YK	Yugen Kaisha	Private
	GK	Gomei Kaisha / Goshi Kaisha	Partnership
Netherlands	NV	Naamloze Vennootschap	Public
	BV	Besloten Vennootschap	Private
Norway	A/S	Aksjeselaskap	Public
Portugal	SA	Sociedada Anonima	Public
	Lda	Limitada	Private
Singapore	Ltd	Limited	Public
	Pty Ltd	Proprietary Limited	Private
South Africa	Pty	Proprietary	Private
Spain	SA	Sociedad Anonima	Public
	Srl	Sociedad Responsabilidad Limitada	Private
Sweden	AB	Aktiebolag	Public
Switzerland	AG	Aktiengesellschaft	Public
	GmbH	Gesellschaft mit Beschrankter Haftung	Private
Switzerland	SA	Societe Anonyme	Public
	Sarl	Societe a Responsabilite Limitee	Private
United Kingdom	plc	Public Limited Company	Public
	Ltd	Limited	Private
United States	Co.	Company	Public/Private
	Corp.	Corporation	Public/Private
	Inc.	Incorporated	Public/Private
Venezuela	CA	Compania Anonima	Public
	SA	Sociedad Anonima	Public
	CRL	Compania de Responsabilidad Limitada	Private
	SRL	Sociedad de Responsabilidad Limitada	Private

APPENDIX 14

US STATES & ABBREVIATIONS

AK	Alaska
AL	Alabama
AR	Arkansas
AZ	Arizona
CA	California
CO	Colorado
CT	Connecticut
DC	District of Colombia
DE	Delaware
FL	Florida
GA	Georgia
HI	Hawaii
IA	Iowa
ID	Idaho
IL	Illinois
IN	Indiana
KS	Kansas
KY	Kentucky
LA	Louisiana
MA	Massachusetts
MD	Maryland
ME	Maine
MI	Michigan
MN	Minnesota
MO	Missouri
MS	Mississippi
MT	Montana
NC	North Carolina
ND	North Dakota
NE	Nebraska
NH	New Hampshire
NJ	New Jersey
NM	New Mexico
NV	Nevada
NY	New York
OH	Ohio
OK	Oklahoma
OR	Oregon
PA	Pennsylvania
RI	Rhode Island
SC	South Carolina
SD	South Dakota
TN	Tennessee
TX	Texas
UT	Utah
VA	Virginia
VT	Vermont
WA	Washington
WI	Wisconsin
WV	West Virginia
WY	Wyoming

APPENDIX 15

MEMBERSHIP OF INTERNATIONAL ORGANIZATIONS

ASSOCIATION OF SOUTH EAST ASIAN NATIONS (ASEAN)

Brunei Darussalam
Indonesia
Malaysia
Philippines
Singapore
Thailand

COMMONWEALTH OF INDEPENDENT STATES (CIS)

Armenia
Azerbaijan
Belarus
Kazakhstan
Kyrgyzstan
Moldova
Russia
Tajikistan
Turkmenistan
Ukraine
Uzbekistan

EUROPEAN ECONOMIC COMMUNITY (EEC)

Belgium
Denmark
France
Germany
Greece
Ireland
Italy
Luxembourg
Netherlands
Portugal
Spain
United Kingdom

EUROPEAN FREE TRADE ASSOCIATION (EFTA)

Austria
Finland
Iceland
Liechtenstein
Norway
Sweden
Switzerland

EUROPEAN SPACE AGENCY (ESA)

Austria
Belgium
Denmark
France
Germany
Ireland
Italy
Netherlands
Norway
Spain
Sweden
Switzerland
United Kingdom

GENERAL AGREEMENT ON TARIFFS & TRADE (GATT)

Antigua & Barbuda
Argentina
Australia
Austria
Bangladesh
Barbados
Belgium
Belize
Benin
Bolivia
Botswana
Brazil
Burkina Faso
Burundi
Cameroon
Canada
Central African Rep.
Chad
Chile
Colombia
Congo
Costa Rica
Cote d'Ivoire
Cuba
Cyprus
Czechoslovakia[1]
Denmark
Dominican Rep.
Egypt
El Salvador
Finland
France
Gabon
Gambia
Germany
Ghana
Greece
Guatemala
Guyana
Haiti
Hong Kong
Hungary
Iceland
India
Indonesia
Ireland
Israel
Italy
Jamaica
Japan
Kenya
Korea, South
Kuwait
Lesotho
Luxembourg
Macau
Madagascar
Malawi
Malaysia
Maldives
Mali
Malta
Mauritania
Mauritius
Mexico
Morocco
Mozambique
Myanmar
Namibia
Netherlands
New Zealand
Nicaragua
Niger
Nigeria
Norway
Pakistan
Peru
Philippines
Poland
Portugal
Romania
Rwanda
Senegal
Sierra Leone
Singapore
South Africa
Spain
Sri Lanka
Suriname
Sweden
Switzerland
Tanzania
Thailand
Togo
Trinidad & Tobago
Tunisia
Turkey
Uganda
United Kingdom
United States
Uruguay
Venezuela
Yugoslavia
Zaire
Zambia
Zimbabwe

[1] Czech Republic and Slovakian Republic from January 1993

INTERNATIONAL ATOMIC ENERGY AGENCY (IAEA)

Afghanistan
Albania
Algeria
Argentina
Australia
Austria
Bangladesh
Belarus
Belgium
Bolivia
Brazil
Bulgaria
Cameroon
Canada
Chile
China, People's Rep.
Colombia
Costa Rica
Cote d'Ivoire
Cuba
Cyprus
Denmark
Dominican Rep.
Ecuador
Egypt
El Salvador
Estonia

INTERNATIONAL ATOMIC ENERGY AGENCY (IAEA) (continued)

Ethiopia
Finland
France
Gabon
Germany
Ghana
Greece
Guatemala
Haiti
Hungary
Iceland
India
Indonesia
Iran
Iraq
Ireland
Israel
Italy
Jamaica
Japan
Jordan
Kampuchea, Democratic
Kenya
Korea, North
Korea, South
Kuwait
Lebanon
Liberia
Libya
Liechtenstein
Luxembourg
Madagascar
Malaysia
Mali
Mauritius
Mexico
Monaco
Mongolia
Morocco
Myanmar
Namibia
Netherlands
New Zealand
Nicaragua
Niger
Nigeria
Norway
Pakistan
Panama
Paraguay
Peru
Philippines
Poland
Qatar
Romania
Russia
Saudi Arabia
Senegal
Sierra Leone
Singapore
Slovenia
South Africa
Spain
Sri Lanka
Sudan
Sweden
Switzerland
Syria
Tanzania
Thailand
Tunisia
Turkey
Uganda
Ukraine
United Arab Emirates
United Kingdom
United States
Uruguay
Venezuela
Vietnam
Yugoslavia
Zaire
Zambia
Zimbabwe

LEAGUE OF ARAB STATES

Algeria
Bahrain
Djibouti
Egypt
Iraq
Jordan
Kuwait
Lebanon
Libya
Mauritania
Morocco
Oman
Palestine Liberation Organization
Qatar
Saudi Arabia
Somalia
Sudan
Syria
Tunisia
United Arab Emirates
Yemen

NORTH ATLANTIC TREATY ORGANIZATION (NATO)

Belgium
Canada
Denmark
France
Germany
Greece
Iceland
Italy
Luxembourg
Netherlands
Norway
Portugal
Spain
Turkey
United Kingdom
United States

ORGANIZATION OF AMERICAN STATES (OAS)

Antigua & Barbuda
Argentina
Bahamas
Barbados
Belize
Bolivia
Brazil
Canada
Chile
Colombia
Costa Rica
Cuba
Dominica
Dominican Republic
Ecuador
El Salvador
Grenada
Guatemala
Guyana
Haiti
Honduras
Jamaica
Mexico
Nicaragua
Panama
Paraguay
Peru
St Kitts & Nevis
St Lucia
St Vincent & Grenadines
Suriname
Trinidad & Tobago
United States
Uruguay
Venezuela

ORGANIZATION OF PETROLEUM EXPORTING COUNTRIES (OPEC)

Algeria
Ecuador
Gabon
Indonesia
Iran
Iraq
Libya
Nigeria
Qatar
Saudi Arabia
United Arab Emirates
Venezuela

UNITED NATIONS (UN)

Afghanistan
Albania
Algeria
Angola
Antigua & Barbuda
Argentina
Armenia
Australia
Austria
Azerbaijan
Bahamas
Bahrain
Bangladesh
Barbados
Belarus
Belgium
Belize
Benin
Bhutan
Bolivia
Bosnia & Herzegovina
Botswana
Brazil
Brunei Darussalam
Bulgaria
Burkina Faso
Burundi
Cambodia
Cameroon
Canada
Cape Verde
Central African Rep.
Chad
Chile
China *
Colombia
Comoros
Congo
Costa Rica
Cote d'Ivoire
Croatia
Cuba
Cyprus
Czech Republic
Denmark
Djibouti
Dominica
Dominican Republic
Ecuador
Egypt
El Salvador
Equatorial Guinea
Estonia
Ethiopia
Fiji
Finland
France *
Gabon
Gambia
Georgia
Germany
Ghana
Greece
Grenada
Guatemala
Guinea
Guinea-Bissau
Guyana
Haiti
Honduras
Hungary
Iceland
India
Indonesia
Iran

UNITED NATIONS (UN) (continued)

Iraq
Ireland
Israel
Italy
Jamaica
Japan
Jordan
Kazakhstan
Kenya
Korea, North
Korea, South
Kuwait
Kyrgyzstan
Laos
Latvia
Lebanon
Lesotho
Liberia
Libya
Liechtenstein
Lithuania
Luxembourg
Madagascar
Malawi
Malaysia
Maldives
Mali
Malta
Marshall Islands
Mauritania
Mauritius
Mexico
Micronesia
Moldova
Mongolia
Morocco
Mozambique
Myanmar
Namibia
Nepal
Netherlands
New Zealand
Nicaragua
Niger
Nigeria
Norway
Oman
Pakistan
Panama
Papua New Guinea
Paraguay
Peru
Philippines
Poland
Portugal
Qatar
Romania
Russia *
Rwanda
Samoa
San Marino
Sao Tome & Principe
Saudi Arabia
Senegal
Seychelles
Sierra Leone
Singapore
Slovak Republic
Slovenia
Solomon Islands
Somalia
South Africa
Spain
Sri Lanka
St Kitts & Nevis
St Lucia
St Vincent & Grenadine
Sudan
Suriname
Swaziland
Sweden
Syria
Tajikistan
Tanzania
Thailand
Togo
Trinidad & Tobago
Tunisia
Turkey
Turkmenistan
Uganda
Ukraine
United Arab Emirates
United Kingdom *
United States *
Uruguay
Uzbekistan
Vanuatu
Venezuela
Vietnam
Yemen
Yugoslavia
Zaire
Zambia
Zimbabwe

* Permanent members of the Security Council.

WESTERN EUROPEAN UNION (WEU)

Greece[1]
Belgium
France
Germany
Italy
Luxembourg
Netherlands
Portugal
Spain
United Kingdom

[1] Membership of WEU awaiting government approval

APPENDIX 16

COUNTRY INFORMATION

COUNTRY	CAPITAL	AREA (sq km)	POP. (thousand)	GNP 1991 ($ million)	(Note 1) DEFENCE SPENDING ($ million)
Afghanistan	Kabul	652090	16560		302
Albania	Tirana	28748	3303		149
Algeria	Algiers	2381741	25798	52239	965
Andorra	Andorra-la-Vella	453	50		
Angola	Luanda	1246700	10301		1013
Anguilla	The Valley	155	7		
Antigua & Barbuda	St John's	280	80	355	
Argentina	Buenos Aires	2780092	32646	91211	856
Armenia	Yerevan	29800	3360	7233	
Aruba	Oranjestad	193	60		
Australia	Canberra	7682300	17341	287765	7599
Austria	Vienna	83857	7730	157528	1593
Azerbaijan	Baku	86600	7219	12065	
Bahamas	Nassau	13864	259	3044	64
Bahrain	Manama	688	518	3679	194
Bangladesh	Dhaka	143998	108756	23449	321
Barbados	Bridgetown	430	258	1711	
Belarus	Minsk	207600	10328	32131	
Belgium	Brussels	30518	9968	192370	3064
Belize	Belmopan	22963	193	389	10
Benin	Porto-Novo	112622	4883	1848	22
Bermuda	Hamilton	53	58		
Bhutan	Thimphu	46500	1467	260	
Bolivia	Sucre	1098581	7356	4799	200
Botswana	Gaborone	581730	1289	3335	136
Brazil	Brasilia	8511996	153164	447324	1093
Brunei Darussalam	Bandar Seri Begawan	5765	264		295
Bulgaria	Sofia	110994	8798	16316	2019
Burkina Faso	Ougadougou	274122	9271	3213	103
Burundi	Bujumbura	27834	5600	1210	32
Cambodia	Phnom Penh	181035	8660	1725	
Cameroon	Yaounde	475442	12081	11320	203
Canada	Ottawa	9215430	26756	568765	11316
Cape Verde	Praia	4033	383	285	10
Cayman Isl.	George Town	260			
Central African Rep.	Bangui	622436	3113	1218	23
Chad	N'djamena	1284000	5828	1212	65
Chile	Santiago	736905	13360	28897	639
China, People's Rep.	Beijing	9572900	1150091	424012	8261
Colombia	Bogota	1141748	32873	41922	959
Comoros	Moroni	1862	492	245	
Congo	Brazzaville	341821	2351	2623	118
Costa Rica	San Jose	51100	2875	6156	74
Cote d'Ivoire	Abidjan	322463	12331	8523	146
Croatia	Zagreb	56538			
Cuba	Havana	110860	10712	1350	
Cyprus	Nicosia	9251	708	6135	190
Czech Republic	Prague	78864	10300		
Czechoslovakia (Note 2)	Prague	127899	15694	38427	1476
Denmark	Copenhagen	43075	5143	121695	2702

LOCAL CURRENCY	CIVIL AIRCRAFT MARKING	INT'L VEHICLE MARKING	INT'L DIALLING CODE	TIME ZONE
Afghani	YA			GMT +4.5
Lek	ZA	AL	355	GMT +1
Dinar	7T	DZ	213	GMT +1
Peseta/Franc	C3	AND	33 628	GMT +1
Kwanza	D2		244	GMT +1
Dollar	VP-LA		1 809	GMT -4
Dollar	V2		1 809	GMT -4
Austral	LV	RA	54	GMT -3
Florin	P4		2 978	GMT -4
Dollar	VH	AUS	61	GMT +10
Schilling	OE		43	GMT +1
Dollar	C6	BS	1 809	GMT -5
Dinar	A9C	BRN	973	GMT +3
Taka	S2		880	GMT +6
Dollar	8P	BDS	1 809	GMT -4
Franc	OO	B	32	GMT +1
Dollar	V3		501	GMT -6
Franc	TY		229	GMT +1
Dollar	VR-B		1 809	GMT -4
Ngultrum	A5		975	GMT +6
Boliviano	CP		591	GMT -4
Pula	A2	RB	267	GMT +2
Cruzeiro	PP/PT	BR	55	GMT -3
	V8	BRU	673	GMT +8
	LZ	BG	359	GMT +2
Franc	XT		226	GMT
Franc	9U	RU	257	GMT +2
				GMT +7
Franc	TJ		237	GMT +1
Dollar	C/CF	CDN	1	GMT -5
Escudos	D4		238	GMT -1
	VR-C		1 809	GMT -5
Franc	TL	RCA	236	GMT +1
Franc	TT		235	GMT +1
Peso	CC	RCH	56	GMT -4
Yuan	B		86	GMT +8
Peso	HK	CO	57	GMT -5
Franc	D6			GMT +3
Franc	TN	RCB	242	GMT +1
Colones	TI	CR	506	GMT -6
Franc			225	GMT
		RC		
	CU	C	53	GMT -5
Pound	5B	CY	357	GMT +2
Koruna			42	GMT +1
Koruna	OK		42	GMT +1
Kroner	OY	DK	45	GMT +1

COUNTRY	CAPITAL	AREA (sq km)	POP. (thousand)	GNP 1991 ($ million)	(Note 1) DEFENCE SPENDING ($ million)
Djibouti	Djibouti	23200	441		
Dominica	Roseau	751	72	175	
Dominican Republic	Santo Domingo	48442	7197	6807	58
Ecuador	Quito	270670	10503	10772	356
Egypt	Cairo	1002000	53087	33068	1650
El Salvador	San Salvador	21041	5308	5697	127
Equatorial Guinea	Malabo	28051	426	142	
Estonia	Tallinn	45100	1591	6088	
Ethiopia	Addis Ababa	1221900	52892	6144	410
Falkland Islands	Stanley	4700			
Fiji	Suva	18333	751	1377	27
Finland	Helsinki	304623	4999	121982	2330
France	Paris	543965	56681	1167749	35728
Gabon	Libreville	267667	1168	4419	138
Gambia	Banjul	11295	901	322	4
Georgia	Tbilisi	69700	5478	9000	
Germany	Berlin	356958	79632	1516785	32556
Ghana	Accra	92456	15336	6176	48
Greece	Athens	131957	10083	65504	4903
Grenada	St George's	344	91	198	
Guatemala	Guatemala City	108889	9466	8816	75
Guinea	Conakry	245857	5873	2669	10
Guinea-Bissau	Bissau	36125	999	194	5
Guyana	Georgetown	83000	802	233	42
Haiti	Port-au-Prince	27750	6603	2471	12
Honduras	Tegucigalpa	112088	5259	3010	127
Hungary	Budapest	93032	10500	28244	861
Iceland	Reykjavik	103000	258	5814	
India	New Delhi	3165596	865020	284668	9626
Indonesia	Jakarta	1919443	181388	111409	1636
Iran	Tehran	1648000	57764	127366	2064
Iraq	Baghdad	438317	19567	6600	
Ireland (Eire)	Dublin	68895	3502	37738	553
Israel	Jerusalem	20770	4888	59128	6550
Italy	Rome	301268	57719	1072198	20421
Jamaica	Kingston	11425	2440	3365	35
Japan	Tokyo	377727	123969	3337191	30405
Jordan	Amman	89206	3453	3881	587
Kazakhstan	Alma-Ata	2717300	16899	41691	
Kenya	Nairobi	582646	25016	8505	259
Kiribati	Tarawa	717	71	53	
Korea, North	Pyongyang	120538	21947	5462	
Korea, South	Seoul	99263	43177	274464	11162
Kuwait	Kuwait	6880	2212	12623	
Kyrgyzstan	Bishkek	198500	4448	6900	
Laos	Vientiane	236800	4279	965	20
Latvia	Riga	63700	2693	9193	
Lebanon	Beirut	10452	2828	120	
Lesotho	Maseru	30355	1816	1053	40
Liberia	Monrovia	111370	2639		

LOCAL CURRENCY	CIVIL AIRCRAFT MARKING	INT'L VEHICLE MARKING	INT'L DIALLING CODE	TIME ZONE
Franc	J2		253	GMT +3
Dollar	J7		1 809	GMT -4
Peso	HI	DOM	1 809	GMT -4
Sucres	HC	EC	593	GMT -5
Pound	SU	ET	20	GMT +2
Colones	YS		503	GMT -6
Bipkwele	3C		240	GMT +1
	ES			
Birr	ET		251	GMT +3
	VP-F		500	GMT -4
Dollar	DQ	FJI	679	GMT +12
Markka	OH	SF	358	GMT +2
Franc	F	F	33	GMT +1
Franc	TR		241	GMT +1
Dalasis	C5	WAG	220	GMT
Deutsche Mark	D	D	49	GMT +1
Cedis	9G	GH	233	GMT
Drachma	SX	GR	30	GMT +2
Dollar	J3	WG	1 809	GMT -4
Quetzales	TG	GCA	502	GMT -6
		3X	224	GMT
		J5	245	GMT
Dollar	8R	GUY	592	GMT -3
Gourdes	HH	RH	509	GMT -5
Lempiras	HR		504	GMT -6
Forint	HA	H	36	GMT +1
Kronur	TF	IS	354	GMT
Rupee	VT	IND	91	GMT +5.5
Rupiah	PK	RI	62	GMT +7
Rial	EP	IR	98	GMT +3.5
Dinar	YI	IRQ	964	GMT +3
Punt	EI	IRL	353	GMT
New Sheqalim	4X	IL	972	GMT +2
Lira	I	I	39	GMT +1
Dollar	6Y	JA	1 809	GMT -5
Yen	JA	J	81	GMT +9
Dinar	JY	HKJ	962	GMT +2
Shilling	5Y	EAK	254	GMT +3
	T3		686	GMT +12
	P		850	GMT +9
Won	HL	ROK	82	GMT +9
Dinar	9K	KWT	965	GMT +3
	RDPL	LAO		GMT +7
	YL			
Pound	OD	RL	961	GMT +2
Loti	7P	LS	266	GMT +2
Dollar	EL	LB	231	GMT

COUNTRY	CAPITAL	AREA (sq km)	POP. (thousand)	GNP 1991 ($ million)	(Note 1) DEFENCE SPENDING ($ million)
Libya	Tripoli	1759540	4714	1922	
Liechtenstein	Vaduz	160			
Lithuania	Vilnius	65200	3765	10220	
Luxembourg	Luxembourg	2586	378	11761	90
Madagascar	Antananarivo	587041	12016	2560	39
Malawi	Lilongwe	94082	8796	1996	19
Malaysia	Kuala Lumpur	329758	18294	45787	1749
Maldives	Male	298	221	101	
Mali	Bamako	1240192	8706	2412	64
Malta	Valletta	246	356	2598	24
Marshall Islands	Majuro	181	48		
Mauritania	Nouakchott	1030700	2023	1026	
Mauritius	Port Louis	2040	1083	2623	
Mexico	Mexico City	1958201	87821	252381	705
Micronesia	Kolonia	701	102		
Moldova	Kishinev	33700	4384	9529	
Monaco	Monaco	2			
Mongolia	Ulan Bator	1566500	2184	350	
Montserrat	Plymouth	106			
Morocco	Rabat	458730	25731	26451	1590
Mozambique	Maputo	799380	16142	1163	120
Myanmar (Burma)	Rangoon	676577	42528	1080	
Namibia	Windhoek	823145	1834	2051	47
Nepal	Kathmandu	140797	19406	3453	42
Netherlands	Amsterdam	41547	15023	278839	7863
New Zealand	Wellington	265018	3429	41626	812
Nicaragua	Managua	130682	3975	1897	79
Niger	Niamey	1267000	7909	2361	49
Nigeria	Abuja	923773	118811	34057	301
Norway	Oslo	323878	4259	102885	3583
Oman	Muscat	300000	1618	8787	1438
Pakistan	Islamabad	796095	115588	46725	3269
Panama	Panama City	77082	2460	5254	75
Papua New Guinea	Port Moresby	462840	4013	3307	40
Paraguay	Asuncion	406752	4441	5374	60
Peru	Lima	1285216	22135	38295	644
Philippines	Manila	300000	62687	46138	1116
Poland	Warsaw	312683	38337	70640	2124
Portugal	Lisbon	91985	10393	58451	1421
Qatar	Doha	11437	452	6968	934
Romania	Bucharest	237500	23276	31079	1304
Russia	Moscow	17075000	148030	479546	
Rwanda	Kigali	26338	7403	1930	40
San Marino	San Marino	61	20		
Sao Tome & Principe	Sao Tome	1000	120	42	
Saudi Arabia	Riyadh	2200000	15431	105133	26809
Senegal	Dakar	196192	7632	5500	117
Seychelles	Victoria	455	69	350	16
Sierra Leone	Freetown	73326	4239	904	4
Singapore	Singapore City	626	3045	39249	2030

LOCAL CURRENCY	CIVIL AIRCRAFT MARKING	INT'L VEHICLE MARKING	INT'L DIALLING CODE	TIME ZONE
Dinar	5A	LAR	218	GMT +1
	HB	FL	41 75	GMT +1
		LY		
Franc	LX	L	352	GMT +1
Franc	RM		261	GMT +3
Kwacha	7Q	MV	265	GMT +2
Ringgit	9M	MAL	60	GMT +8
Rufiyaa	8Q		960	GMT +5
Franc	TZ	RMM	223	GMT
Lira	9H	M	356	GMT +1
	MI		692	GMT +12
Ouguiyas	5T	RIM	222	GMT
Rupee	3B	MS	230	GMT +4
Peso	XA/XB/XC	MEX	52	GMT -6
			691	GMT +11
	3A	MC	33 93	GMT +1
		BNMAU		GMT +8
		VP-LM	1 809	GMT -4
Dirhams	CN	MA	212	GMT
	C9		258	GMT +2
Kyat	XY		95	GMT +6.5
	V5		264	GMT +2
Rupee	9N		977	GMT +5.75
Guilder	PH	NL	31	GMT +1
Dollar	ZK	NZ	64	GMT +12
Cordoba	YN		505	GMT -6
Franc	5U	NIG	227	GMT +1
Naira	5N	WAN	234	GMT +1
Kroner	LN	N	47	GMT +1
Rial	A40		968	GMT +4
Rupee	AP	PAK	92	GMT +5
Balboa	HP	PA	507	GMT -5
Kina	P2		675	GMT +10
Guaranies	ZP	PY	595	GMT -4
New Soles	OB	PE	51	GMT -5
Peso	RP	PI	63	GMT +8
Zloty	SP	PL	48	GMT +1
Escudos	CS	P	351	GMT
Riyal	A7		974	GMT +3
Lei	YR	R	40	GMT +2
			7	GMT +3
Franc	9XR	RWA	250	GMT +2
	T7	RSM	39 549	GMT +1
		S9	239	GMT
Riyal	HZ		966	GMT +3
Franc	6V	SN	221	GMT
Rupee	S7	SY	248	GMT +4
Leones	9L	WAL	232	GMT
Dollar	9V	SGP	65	GMT +8

COUNTRY	CAPITAL	AREA (sq km)	POP. (thousand)	GNP 1991 ($ million)	(Note 1) DEFENCE SPENDING ($ million)
Slovakian Republic	Bratislava	49035	5260		
Slovenia	Ljubljana	20251			
Solomon Islands	Honiara	27556	326	184	
Somalia	Mogadishu	637657	8041		
South Africa	Pretoria	1127200	36762	90953	4106
Spain	Madrid	492592	39045	486614	8418
Sri Lanka	Colombo	65610	17194	8665	465
St Helena	Jamestown	122	290		
St Kitts & Nevis	Basseterre	262	39	156	
St Lucia	Castries	617	152	380	
St Vincent & Grenadines	Kingstown	388	108	187	
Sudan	Khartoum	2505813	25855	10107	350
Suriname	Paramaribo	163820	457	1649	65
Swaziland	Mbabane	17400	825	874	
Sweden	Stockholm	449964	8588	218934	5765
Switzerland	Berne	41293	6740	225890	4102
Syria	Damascus	185180	12824	14234	1754
Tajikistan	Dushanbe	143100	5412	5669	
Tanzania	Dodoma	945037	25270	2424	12
Thailand	Bangkok	513115	56679	89548	2686
Togo	Lome	56785	3761	1530	25
Tonga	Nuku'alofa	748	100	110	
Trinidad & Tobago	Port-of-Spain	5124	1249	4525	
Tunisia	Tunis	164150	8223	12417	459
Turkey	Ankara	779452	57237	103888	5692
Turkmenistan	Ashkhabad	488100	3748	6387	
Turks & Caicos Isl.	Grand Turk	430			
Tuvalu	Fongafale	24			
Uganda	Kampala	241038	16876	2762	76
Ukraine	Kiev	603700	51999	121458	
United Arab Emirates	Abu Dhabi	83657	1630	32813	4900
United Kingdom	London	229880	57536	963696	40692
United States	Washington DC	9372614	252040	5686038	306422
Uruguay	Montevideo	176215	3110	8895	190
Uzbekistan	Tashkent	447400	20955	28255	
Venezuela	Caracas	912050	20191	52775	2077
Vietnam	Hanoi	330363	67843	2001	
W. Samoa	Apia	2831	168	156	
Yemen	Sana'a	531000	12533	6746	1058
Yugoslavia (Note 3)	Belgrade	255804	23690	70038	4145
Zaire	Kinshasa	2344885	38473	8123	20
Zambia	Lusaka	752614	8373	3394	96
Zimbabwe	Harare	390759	10080	6220	361

Note 1. Estimated 1991 defence expenditure in 1990 constant dollars.
Note 2. Czech Republic and Slovakian Republic from January 1993 (please refer)
Note 3. Refers to the former Yugoslavia.

LOCAL CURRENCY	CIVIL AIRCRAFT MARKING	INT'L VEHICLE MARKING	INT'L DIALLING CODE	TIME ZONE
Koruna			42	GMT +1
	SL			
Dollar	H4		677	GMT +11
Shilling	6O		252	GMT +3
Rand	ZS	ZA	27	GMT +2
Peseta	EC	E	34	GMT +1
Rupee	4R	CL	94	GMT +5.5
				GMT
Dollar	V4		1 809	GMT -4
Dollar	J6	WL	1 809	GMT -4
Dollar	J8	WG	1 809	GMT -4
Pound	ST			GMT +2
Guilder	PZ	SME	597	GMT -3
Lilangeni	3D	SD	268	GMT +2
Kronor	SE	S	46	GMT +1
Franc	HB	CH	41	GMT +1
Pound	YK	SYR	963	GMT +2
Shilling	5H	EAT	255	GMT +3
Baht	HS		66	GMT +7
Franc	5V	TG	228	GMT
Pa'anga	A3		676	GMT +13
Dollar	9Y	TT	1 809	GMT -4
Dinar	TS	TN	216	GMT +1
Lira	TC	TR	90	GMT +2
	VQ-T		1 809	GMT -5
	T2		688	GMT +12
Shilling	5X	EAU	256	GMT +3
Dirham	A6		971	GMT +4
Pound	G		44	GMT
Dollar	N	USA	1	GMT -5
New Peso	CX		598	GMT -3
Bolivares	YV	YV	58	GMT -4
	VN	VN	84	GMT +7
Tala	5W	WS	685	GMT -11
Rial	7O		967	GMT +3
Dinar	YU	YU	38	GMT +1
Zaires	9Q/9T	ZR	243	GMT +1
Kwacha	9J	Z	260	GMT +2
Dollar	Z	ZW	263	GMT +2